A Guide to the Climate Debate

The science and politics of climate change

Richard Cox

The only true wisdom is in knowing you know nothing
(Socrates: Greek Philosopher, 470 BC-399 BC)

Copyright © Richard Cox
September 2024

KDP ISBN: 9798338201992

A Guide to the Climate Debate

Contents

List of figures	4
About the author	6
Acknowledgements	7
Forword	8
Preamble	11
Key questions:	16
Is the earth warming?	17
Has there been an increase in extreme weather?	23
What is the natural carbon cycle and its relevance?	29
Are rising CO_2 levels due to human activity?	36
Does rising CO_2 concentration cause global warming?	40
How does the sun influence the climate?	48
Does rising CO_2 concentration lead or lag warming?	52
What about sea level rise and ocean acidification?	57
What is the record of past climate change predictions?	66
Why has climate change become political?	86

A Guide to the Climate Debate

What are the implications of Net Zero policies?	96
What are the alternatives to Net Zero?	112
Findings and conclusions	126
Appendices:	
Glossary	139
References	144
Reading list	150

Usage notes:
- Web links (URLs) are provided for the reader's convenience throughout this document identified by the format, 'Web link example' [Ref X]. Best endeavours have been used to ensure these URLs to 3rd party sites are correct and active at the time of writing this article. We make no guarantee that these websites will remain active.
- The information obtained from these web sites and referenced documents is available in the public domain.
- A glossary is also provided to help with the inevitable climate science lingo.

A Guide to the Climate Debate

List of figures

Fig
1	Central England Mean Temperature 1659 to 2022	19
2	Global temperatures from 1880	21
3	US Annual Heat Wave Index, 1895 - 2021	24
4	Precipitation Worldwide, 1901 - 2021	26
5	Cumulative area burnt by wildfires by week, World	28
6	Simplified schematic of the global carbon cycle…	30
7	Relationship between tomato canopy photosynthesis and CO_2…	32
8	Ocean circulations…	35
9	CO_2 levels over the past 500 million years	36
10	Temperature estimates over the past 500 million years	37
11	Global Carbon Dioxide over 800,000 Years	38
12	Global atmospheric CO_2 compared to annual emissions…	39
13	The Greenhouse Effect	40
14	Global energy flows	42
15	Arctic surface air temperature compared with total solar…	44
16	Global average temperature and measurement of the sun's…	46

17	The earth's orbit around the Sun, showing the key parameters...	49
18	Temperature and CO_2...	55
19	Global Sea Level changes	59
20	Infamous hockey stick curve reproduced by the IPCC...	71
21	Mann at the centre of a paleoclimate web	79
22	AR4 Figure SPM.1 shows the smoothed surface temperature...	82
23	Evolution of CO_2 energy emission...	104
24	Presence of CO_2 in the atmosphere	105
25	Global CO_2 emissions from fossil fuels and industry	107
26	Annual CO_2 emissions by world region.....	108
27	Sustainable Development Goals	120

About the author

As a child **Richard Cox** expressed an unusual mix of interests and career aspirations for:

- Wildlife and wild places, with a desire to be a naturalist and explorer, part today's Sir David Attenborough and part Sir Ranulph Fiennes.
- Science, engineering and electricity, having survived several mains voltage electric shocks during unsupervised home experiments.
- To become a police detective, in order to solve crimes and to bring the perpetrators to justice.

Electrical engineering won out and after serving in the Royal Navy (Fleet Air Arm), he spent the remainder of his career in the electrical power generation industry. However, he did achieve an earlier career aspiration by becoming a weekend 'volunteer naturalist' at a California wilderness park during the early 1980s. He considered himself an environmentalist long before the media discovered the word and before the extremists corrupted its meaning. Having travelled widely he has been fortunate to have visited some of the world's wild and interesting places. Now retired he still maintains a keen interest in science and protection of the environment.

Acknowledgments

To **Professor David Unwin**, a retired climatologist, for providing an insight into the academic world of climatology, and the inspiration to embark on this review of evidence into climate change. David would describe himself as a retired academic who taught what back in the day was called 'climatology' at various universities. He helped to set me on my path with this investigation by providing some reference material and general guidance into the science of climatology. David stresses that he is not a 'climate change denier' but freely admits to being a 'greenhouse sceptic'.

To **Julie** for supporting me throughout this lengthy evidence review and writing of this climate guide. Also, for proof reading and providing a sanity check on the final document.

A Guide to the Climate Debate

Foreword

This examination of the science and politics of climate change started in early 2024 and from this author's established belief in anthropogenic climate change. While supporting the widespread acceptance of man-made global warming, he had concerns with some of the alarmist claims, including the claimed scientific consensus.

Regarding the Net Zero policies adopted by governments as the solution to climate change, this author had already become highly sceptical, especially with the associated greenwashing and evidence showing that many so-called green policies would be unlikely to actually provide any CO_2 emission reductions, while having many undesirable environment consequences.

The documentary Climate the Movie [Ref 1] released March 2024 had a profound effect by challenging this author's belief in anthropogenic climate change, and touched on some existing concerns with the science. Prior to this point, a lack of knowledge of climate science prevented any proper examination of the anthropogenic climate change theory, simply through not knowing where to start. Viewing '**Climate the Movie**' helped in the formation of the key questions that needed to be answered and these key questions form the basis of this investigation.

It also became apparent that almost anyone this author spoke to on the subject of climate change or global warming was equally ill informed on the science and simply believed the claims by the experts and the media. As this investigation proceeded, a realisation emerged of the need for a guide to climate change that examined the science in sufficient detail to establish the facts yet remain comprehensible to the ordinary person. '**A guide to the climate debate**' was therefore conceived.

The reader may well ask what qualifies an engineer to write about climate change? While an engineering background would not provide the academic qualification to work in any of the climate sciences, an engineering mindset does provide the analytical skills needed to investigate and assess the scientific claims and associated politics.

It can be argued that what matters most in such an endeavour is having a methodical and investigative mindset alongside an understanding of basic science. An internet search on *'engineers' mindset'* comes up with the following descriptions:

- *"Engineers use a unique mode of thinking based on seeing everything as a system. They see structures that aren't apparent to the layperson"*
- *"Thinking in systems means that you can deconstruct (breaking down a larger system into its modules) and reconstruct (putting it back together)."*
- *"Engineers think analytically. Einstein said, if you can't explain it simply, then you don't understand it well enough."*
- *"We think of a rational, and methodical process."*
- *"Critical thinking and abstract thought"; "Insatiable Curiosity"; "Attention to Detail"*

In many ways there are similarities between what is in effect the global climate control system and engineered control systems such as an aircraft flight control system or a power station steam turbine generator control system. However, the global climate control system is much more complex with an almost infinite number of control parameters.

This guide to the climate debate is aimed at a wide range of reader abilities including those with minimal scientific knowledge. It is also aimed at those who are concerned for the environment and would like to understand the fundamentals of climate science and the implications of climate change.

The references and web links provide further information for those who wish to delve further. This evidence review attempts a balanced appraisal in a logical manner examining the competing claims by those who are generally

A Guide to the Climate Debate

known as the '**Warmists**' sometimes known as Alarmists and the '**Sceptics**' sometimes known as Deniers.

The '**Warmists**' support the climate change hypothesis of man-made global warming, characterised in the media by activist stereotypes such as Greta Thunberg, Just Stop Oil, Extinction Rebellion and some left-wing politicians, supported by much of the mainstream media.

However, the Warmists also include a much larger scientific community where it is claimed there is a 'consensus' on the science supporting the climate change hypothesis. Many Western governments are following this 'climate science' by embracing policies aimed at reducing greenhouse gas emissions.

The '**Sceptics**' are characterised in the media by stereotypes such as Donald Trump, his MAGA supporters, a broad spectrum of so-called 'Climate Deniers', some of the right-wing media and politicians. The Sceptics also include a mostly silent but highly credible scientific community who challenge the scientific aspects of the claimed 'consensus' on climate change.

However, anyone that openly challenges the climate change consensus today, whether scientist, politician or layperson, risks being labelled a '**Denier**' with derogatory overtones of 'Holocaust Denier'.

Sceptic - A scientific or empirical sceptic is one who questions beliefs on the basis of scientific understanding and empirical evidence. Most scientists, being scientific sceptics, test the reliability of certain kinds of claims by subjecting them to systematic investigation via the 'scientific method'.

A Guide to the Climate Debate

Preamble

Awareness of climate change and its potential impacts began as a gradual process for many people in the UK including this author, from the late 1990s onwards. This growing awareness was aided by the Al Gore documentary movie titled 'An Inconvenient Truth' [Ref 2] released in 2006, (still available online on payment of a fee). This movie provided a highly convincing argument in 2006 of the claimed climate change hypothesis and global warming threat. Meanwhile, the 'United Nations' [Ref 3] took up the fight against climate change with its body the 'Intergovernmental Panel on Climate Change' (IPCC) [Ref 4] assessing the science related to climate change and advising governments.

The claimed cause of the global climate emergency is a rise in the concentration of atmospheric greenhouse gases, mainly carbon dioxide (CO_2) and that this is caused by human activities, especially the use of fossil fuels, oil, coal and gas. This rising atmospheric greenhouse gas concentration is in turn causing global temperatures to rise due to the so-called greenhouse effect. The correct term for man-made global warming is *'anthropogenic climate change'*.

Dire warnings were made at that time (2006) when it was predicted that the polar icecaps would have melted by around 2014. For many of us non-climatologists in the UK there was no reason not to believe the experts or their predictions. Many of us accepted that climate change was real, however, it was all predicted for decades into the future so at that time most people were not unduly concerned.

Over the following years there was a growing media focus on global warming accompanied by more convincing arguments and alarmist climate warnings. One of the early predictions was that atmospheric CO_2 concentration and hence global temperatures would continue to rise until a tipping point was reached, resulting in a rapid runaway warming effect making the earth uninhabitable. The only claimed solution being the elimination of man-made

A Guide to the Climate Debate

CO_2 emissions. All this continued to sound both plausible and frightening to most people.

The climate change agenda continued to gain momentum, especially in the UK leading to the passing of the Climate Change Act during 2008. This act of parliament committed the UK to reducing its greenhouse gas emissions to less than 80% of 1990 levels by 2050. This ambitious target was made even more ambitious during 2019 when the UK became the first major economy to commit to a Net Zero target for greenhouse gas emissions by 2050.

Then in 2021 the UK government followed with an even more ambitious policy to fully decarbonise UK power generation by 2035. This means eliminating the use of fossil fuels such as coal, oil and gas. The climate change act and its amendments in 2019 and 2021 occurred with minimal public debate and with most people having no idea of the implications. Then, during July 2024 the new Labour government were elected with a commitment to advancing the decarbonisation of UK power generation to 2030.

It is only now in 2024 with the growing awareness of the impacts of the UK Net Zero policy such as the phasing out of gas and oil for home heating, and petrol and diesel cars, that more people are becoming concerned with the sacrifices that will have to be made. Achieving Net Zero by 2050 or decarbonising UK power generation by 2030 or even 2035 is considered by most informed engineers to be technically impossible to achieve without draconian measures including power rationing, having a significant impact on our economy, freedoms and way of life. Net Zero policy is now a legal requirement on the UK government but its implementation threatens to make the UK a very cold, dark and impoverished place to live.

If the dangers of global warming are as serious as some of the predictions, then making sacrifices now to ensure the survival of the planet for future generations will be fully justified. However, the magnitude of the sacrifices justifies a much more critical look at the science behind the global warming hypothesis, the courses of action being undertaken by the UK and other governments and whether there are more effective alternatives. If we are to make sacrifices then we need to make the right ones. Governments are not renowned for making well thought out decisions and the UK Net Zero policy is no exception.

The many alarmist climate predictions made over the years by claimed climate experts have either not materialised or have fallen short of what was originally predicted including the infamous claim that the polar ice would have melted by 2014. All this has cast some doubt as to the credibility of the climate science. It is also apparent that very few people outside of a small group of climate scientists actually understand this climate science. This review of evidence sets out to examine the climate science, the associated politics and explain it in a way that provides for greater transparency and understanding.

The change in public and government perception to climate change has been brought about through a long-term campaign of alarmist climate education by the media that some may consider propaganda or brain washing. The information provided has regularly made reference to extreme weather and impending doom but with little scientific justification for the claims. While the lack of scientific detail does not mean the science is wrong, it does result in a lower level of confidence, especially for those with an investigative mind, leaving the climate change hypotheses open to challenge. It is also fairly clear to this author that very few if any UK members of parliament who voted for the climate change act and its amendments will have understood the science supporting the UK government's Net Zero policy or its true implications for society.

Unfortunately, to openly question any aspects of the global warming hypothesis in the UK today carries the risk of being labelled a 'Denier,' or worse, and presents a career risk for many occupations.

Meanwhile, outside of the Western world, especially in China, only apparent lip service is being given to reducing greenhouse gas emissions. China currently burns more coal than the rest of the world combined. Most of the reductions in greenhouse gas emissions claimed by the UK have in effect been exported to China and other countries as heavy manufacturing moves out of the UK. It is quite clear that the UK contribution to reducing greenhouse gas emissions will have a negligible effect on the total global emissions as long as China and other countries continue to increase their emissions.

The evolving climate change narrative over the last decades has highlighted a number of concerns for this author, some described here as 'red flag' issues. These concerns include:

- The claimed **'consensus'** with the science of climate change and that **'the science is settled'** is a major red flag issue. Science is never settled, especially on such a complex topic where even today there is still so much regarding the earth's climate that is not understood.
- The apparent purging of scientists who dare to challenge the scientific consensus. The purge of dissent in the UK appears to be complete such that any dissent or 'coming out' on climate change would be career suicide for any climate scientist today. Any remaining dissenting scientists have either been ostracised or are now long retired. This absence of proper scientific checks and balances is another red flag issue.
- Over the last two decades, the nation has been inundated with climate change propaganda but without any real scientific debate. Alternative scientific views have been excluded to give us the science equivalent to a single party State.
- Greenwashing, the practice of making false or misleading claims regarding green credentials by businesses and by government proliferate. An example is the subsidised burning of trees at the Drax power station in the UK, imported from North American hardwood forests including old growth forests. This biomass fired power generation produces more CO_2 emissions than the original burning of coal that it replaced but is classified as 'sustainable' and CO_2 neutral since the trees will regrow, eventually, even though that may take over 100 years. Vast amounts of tax payers' money are being wasted by the government to subsidise this ineffective means of reducing CO_2 emissions. Greenwashing reduces the credibility of government climate policies.

- Over the last two or three decades climate change has moved on from science to an ideology, akin to a religion, now known as 'climatism'. At the same time climate change has become a multi-billion-dollar self-perpetuating global enterprise.
- During the Covid lockdowns the UK government demonstrated how it can arbitrarily control the population in a national emergency. Climate change could easily be considered as a national emergency, so for libertarians this must also be a red flag issue.

A Guide to the Climate Debate

Key questions

The documentary Climate the Movie [Ref 1] makes a number of claims challenging the climate change hypothesis. These claims along with this author's pre-existing concerns resulted in the creation of 12 key questions that form the structure of this evidence review. In this section each of these key questions is examined in order to better understand the science and politics of climate change and to check out the validity of the competing Warmist and Sceptic arguments. Questions 1 to 8 consider the science and questions 9 to 12 consider the associated politics.

1. Is the earth warming?
2. Has there been an increase in extreme weather?
3. What is the natural carbon cycle and its relevance?
4. Are rising CO_2 levels due to human activity?
5. Does rising CO_2 concentration cause global warming?
6. How does the sun influence the climate?
7. Does rising CO_2 concentration lead or lag warming?
8. What about sea level rise and ocean acidification?
9. What is the record of climate change predictions?
10. Why has climate change become political?
11. What are the implications of Net Zero policies?
12. What are the alternatives to Net Zero?

A Guide to the Climate Debate

Q1. Is the earth warming?

First, we need to be clear what is meant by the words weather and climate;

- Weather is the state of the atmosphere at any given time and place. Aspects of the weather include temperature, precipitation, clouds, wind and severe weather conditions include hurricanes, heatwaves, floods and droughts.

- Climate is the long-term average of the weather in a given place and can be defined as *'the sequence of weather elements we have learned to expect'*. Climate is defined not only by average temperature and precipitation but also by the type, frequency, duration and intensity of weather events such as heat waves, cold spells, storms, floods and droughts.

Climate is a concept invented by humans and designed originally to be compatible with a human sized lifespan. Every climate is defined by a timescale and we can change the climate by either lengthening or shortening the timescale. It is interesting in that as late as the 1960s the UK Met Office routinely denied any climate definition on timescales as short as a human lifespan. Today the UK Met Office - What is climate [Ref 5] web page states; *"We usually define a region's climate over a period of 30 years"*, so a significant shortening.

Warmists tend to use shorter climate time periods. Shorter time periods enable a more convincing demonstration of climate change. In contrast the Sceptics tend to use longer climate time periods as this enables a more convincing demonstration using historical data that climate has always been changing, and will likely continue to change due to natural causes. When examined over much longer timespans, up to 1 million years and beyond, it can be interpreted that today's temperatures, CO_2 levels and their fluctuations are not that unusual.

A Guide to the Climate Debate

The earth has experienced many cycles of warm periods (often referred to as a climate optimum) and cold periods (often referred to as an ice age). The Sceptics also claim that global temperatures and CO_2 levels have been higher in the past. It is also claimed by Sceptics that the earth, or at least Europe is currently coming out of the last mini-ice age, hence we are passing through a natural warming period. The earth's climate history as claimed by the Sceptics appears to be supported by the historical climate data.

There is also some disagreement on the accuracy of the data as direct global weather measurements are only available for around the last 150 years creating one of the great challenges for climate science, the lack of historical weather data. What is available are surrogates including analysis of lake sediments, tree rings etc.

These surrogates come with limitations, with tree ring data found to be less reliable an indicator of historical global temperature as is often claimed since other climatic factors can have a major influence on tree growth rates. This is discussed later at key question 9, the Hockey Stick deception, where the misuse of tree ring data was an integral part of the deception.

Possibly the most accurate evidence comes from Antarctic and Greenland ice cores providing data going back around 800,000 years.

The longest running instrumental temperature series in the world is the Central England Temperature (CET) series published by the English climatologist Gordon Manley in 1953, and subsequently extended and updated in 1974. See the History of Information Central England Temperature record [Ref 6] web page.

This temperature record covers the Midlands area of England in degrees Celsius for the time period from 1659. This record is now maintained and updated by the UK Met Office CET [Ref 7] and shows the temperature record for Central England from 1660 to 2022, reproduced here as Figure 1.

A Guide to the Climate Debate

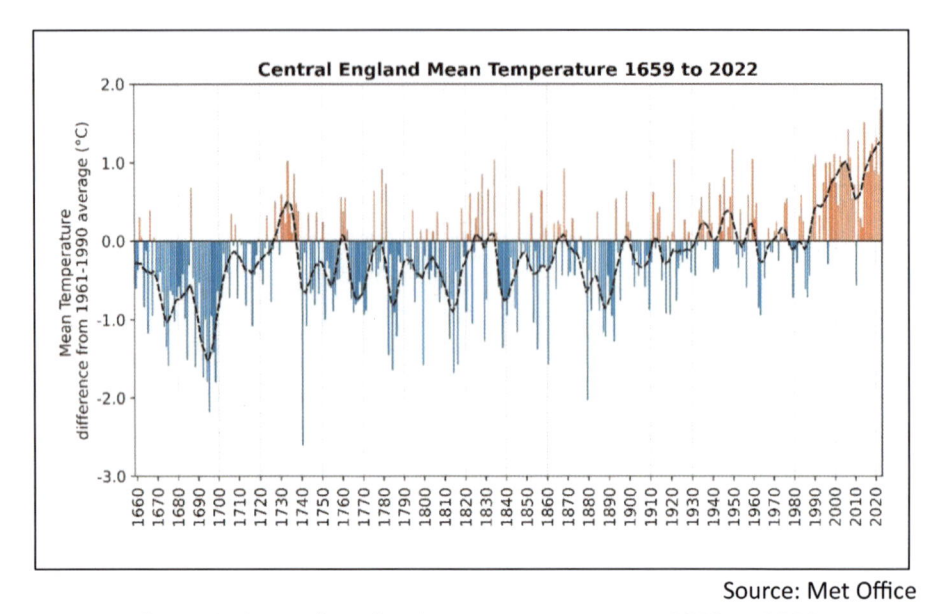

Figure 1: Central England Mean Temperature 1659 to 2022
Source: Met Office

There is however, disagreement on how corrections have been made for the urban heat island (UHI) effect. The UHI effect is the warming of urban areas due to development and human activity compared to rural areas, and has been studied by scientists for many years. The Sceptics claim that the Warmists are still inflating the recent global warming effect due to the use of measurements taken from weather stations that were originally located in rural areas that have since become urbanised, and from applying inadequate UHI corrections.

These temperature measurements are crucial to accurately deriving global temperature. This land temperature measurements data is managed by the Climatic Research Unit (CRU) at the University of East Anglia (UEA). The measurements provided by the CRU and the Met Office are then used in IPCC reports.

It was recognised long ago that insufficient UHI correction could exaggerate the recent warming trend in support of the IPCC climate change hypothesis,

A Guide to the Climate Debate

so has long been an issue for investigation by Sceptics. When we consider that the UHI effect tends be larger than the global temperature rise currently being measured, these corrections are therefore highly significant.

The book **'Hiding the Decline'** [Ref D] by Andrew Montford discusses at Chapter 2 the attempts by Sceptics to verify the temperature data and its analysis by the CRU and to understand how the UHI corrections were made. Montford discusses in some detail the lengths the CRU scientists went to in preventing access to the data and its analysis including obstruction of Freedom of Information (FOI) requests. The importance of these early instrumental temperature records cannot be overstated as these records going back around 150 years are used to establish the baseline for subsequent claims of the global temperature rise since the pre-industrial era.

A discussion on the impact of the UHI effect is provided by David Unwin in his paper titled, Teaching a model based climatology using energy balance simulation [Ref 8] dated 1981. Additionally, Unwin has recently summarised his concerns of the UHI impact, with;

"I do not think that a simple 'correction' in which a number that might as well be drawn from a hat is subtracted from the station temperature record. There are several reasons for this. First, the UHI arises from a number of causes mostly and usually related to the night time energy balance in settled anticyclonic conditions. If the record is 'contaminated' by such incidents the effect is seen in the recorded night time minima and I note with concern that much of the suggested 'global warming' turns out to be in that part of the data. Second, different cities/towns/villages have different rates of urbanisation such that using a surrogate for 'urban' such as population, 'urban' area covered and so on doesn't get you very far. If anything, I'd use some very local measurements at the recording station as the surrogate. Third, I know from my own work that quite small settlements can generate UHI. Whether this contaminates their data significantly is a statistical question that I have yet to see answered. Finally, and somewhat more philosophically, one needs to factor in changes in the weather itself – anything that gives more anticyclonic weather will (at least in the mid latitudes) give more and bigger heat island incidents. Willy Soons' little test using totally

A Guide to the Climate Debate

rural stations that showed no AGW should have been a warning, but it has been quietly ignored."

For this evidence review, several sources of global temperature data have been examined, with the United States Environmental Protection Agency (US EPA) Climate Change Indicators: U.S. and Global Temperature | US EPA [Ref 9] web site providing some useful data analysis.

> We now discover following the Climategate release of emails, (discussed at key question 9) that the raw temperature data has been destroyed by the CRU such that it will never be possible to verify the CRU temperature data or how the UHI corrections were made.

The US EPA Figure 2 shows how surface temperatures have increased over the period 1901 to 2021. The data shows a warming trend since the 1980s and worldwide, 2016 is claimed as the warmest year on record.

Source: NASA earth observatory

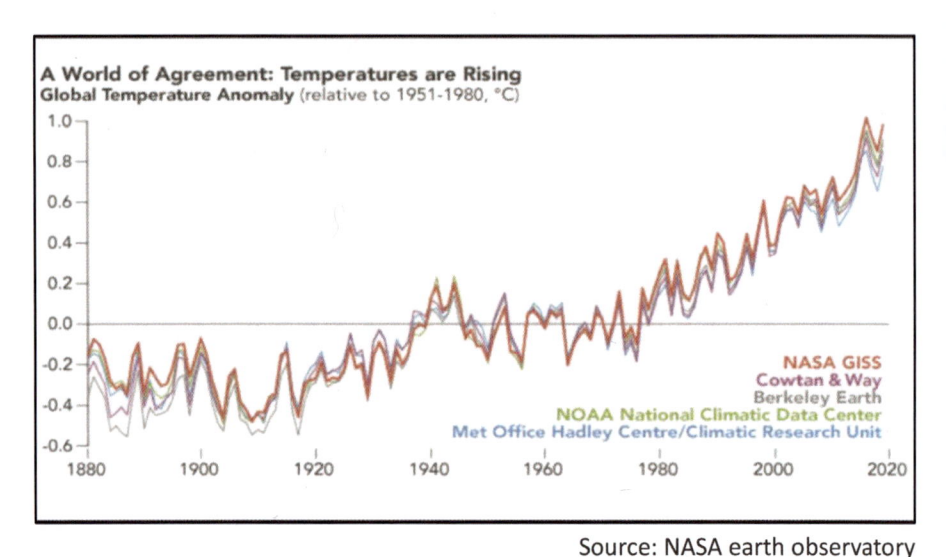

Figure 2: Global temperatures from 1880

A Guide to the Climate Debate

The NASA earth observatory Global temperature [Ref 10] web site also provides evidence confirming this rise in global temperature over the last decades, reproduced here as Figure 2. It is claimed by the United Nations that in 2024 the earth is about 1.2°C warmer than during the pre-industrial era (1850-1900). However, what Figure 2 does not tell us is whether this claimed 1.2°C temperature rise is unusual in a historical context.

If we examine the earth's recent climate history, say the last million or even the last thousand years we find that this temperature increase is not unusual. In his book **'Climate: the great delusion'** [Ref B] Gerondeau informs us on page 84 that during the medieval climate optimum (Medieval Warm Period) around the year 1000, the world was warmer than today, confirmed by the fact that the Vikings settled Greenland for 300 years growing cereal crops and raised cattle.

Despite some dispute of the actual temperature rise, the evidence shows that the climate is changing and that the earth is warming with a claim to be 1.2°C warmer than during the pre-industrial era. While the earth is shown to be warming, climate history tells us the current warming trend is not unusual, it being warmer less than 1,000 years ago during the Medieval Warm Period.

It can also be argued that a 1.2°C global temperature rise, or even the potential for a future 2°C or 3°C rise should not in itself present a significant environmental crisis for the earth or for most of its occupants. However, what may be of concern though are the effects of extreme weather conditions and that is the subject of the next key question.

Q2. Has there been an increase in extreme weather?

In order to assess this, we need to examine historical weather data for the analysis of the different types of extreme events. For a weather event to be considered as extreme there needs to be a long enough historical record to determine how unusual it is and there needs to be someone around to witness and record it. The analysis of extreme weather events is therefore hampered by the lack of historical records.

We should also recognise that extreme weather events vary from year to year and that such events are expected and normal in the absence of climate change. However, as global temperature and hence air temperatures increase it is expected it will lead to an overall increase in global precipitation. In assessing the data, we are looking for long term trends in the frequency and severity of these extreme weather events.

Data and analysis provided by the United States Environmental Protection Agency (US EPA) on their EPA Climate change indicators [Ref 11] web pages provides some useful extreme weather data. Although this data is focused mainly on the United States it helps in the understanding of extreme weather globally and across different climatic zones. The main extreme weather conditions categorised are:

High and low temperatures

The US EPA high & low temperature data [Ref 12] web page describes the trends in unusually hot and cold temperatures across the USA. Figures 1 to 5 shows that unusually hot summer days have become more common over the last 30 years with the unusually hot summer night lows increasing at an even faster rate, indicating less cooling off at night.

A Guide to the Climate Debate

Heat waves

The US [EPA Heat Wave](#) [Ref 13] web page Figure 1 covering the period 1961 to 2021 shows that for the USA the heat wave frequency and heat wave season length have increased significantly. Heat wave frequency has increased from an average of two per year in the 1960s to around six per year in the 2020s. Heat wave duration and intensity have also increased.

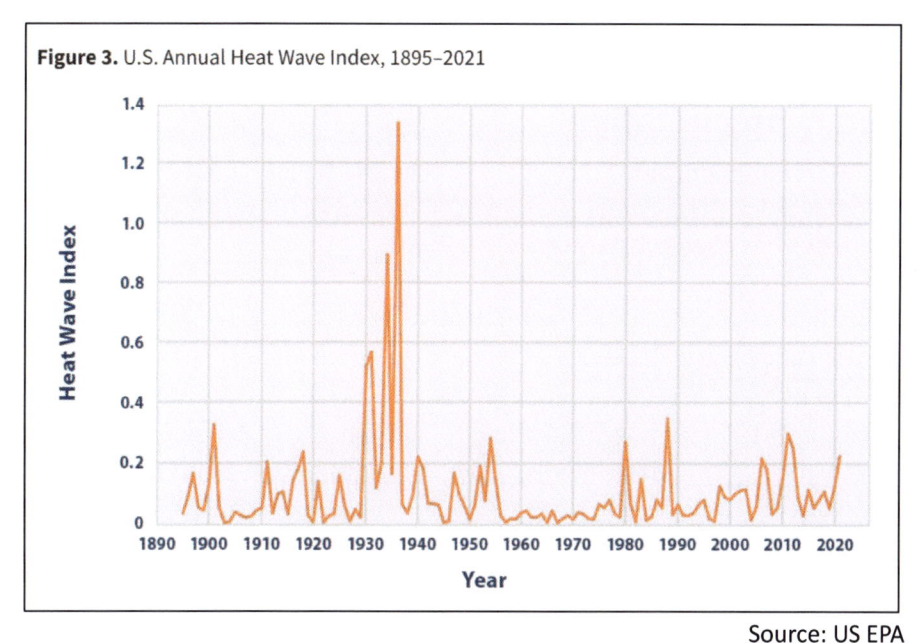

Figure 3. U.S. Annual Heat Wave Index, 1895–2021

Source: US EPA

Figure 3: US Annual Heat Wave Index, 1895 - 2021

US EPA Figure 3, reproduced here as Figure 3 shows the US annual heat wave index over the period 1895 to 2021 with the heat spike in the 1930s. This large spike represents the extreme, persistent heat waves in the Great Plains region.

Drought

The US EPA Drought [Ref 14] web page, Figures 1 to 4 show that average US drought conditions have varied over the 1895 to 2020 time period with a slight trend overall towards wetter conditions. In terms of scale and duration, the droughts of the 1930s 'Dust Bowel' era remain the most extreme in the historical record. US EPA Figure 3 shows that over the period 1900 to 2020 the south western USA has become drier and eastern states have become wetter indicating a changing weather pattern.

Heavy precipitation

The US EPA Heavy precipitation [Ref 15] web page Figures 1 & 2 show that extreme single-day events remained fairly steady between 1910 and the 1980s but has risen substantially since then.

Global precipitation

The US EPA Global Precipitation [Ref 16] web page, Figure 2 reproduced here as Figure 4 shows precipitation worldwide from 1901 to 2021. Global precipitation has increased slightly since 1901 at an average rate of 0.04 inches per decade. If global surface temperature continues to rise then precipitation is also expected to continue to rise.

Flooding

The US EPA River flooding [Ref 17] Figures 1 & 2 show the changes in magnitude and frequency of flooding. Floods have generally become larger in the Northeast and Midwest, and decreased in the West. Floods have become more frequent in the Northeast, Pacific Northwest and northern Great Plains. Flood frequency has decreased in the Southwest and Rockies, again indicating a changing weather pattern. If we look at flooding in the UK, we find that a major contributor is the building on floodplains and neglect of natural and artificial flood defence mechanisms, and little to do with extreme weather or climate change. A useful source of extreme UK weather data is the [TORRO](#) website [Ref 18].

Figure 2. Precipitation Worldwide, 1901–2021

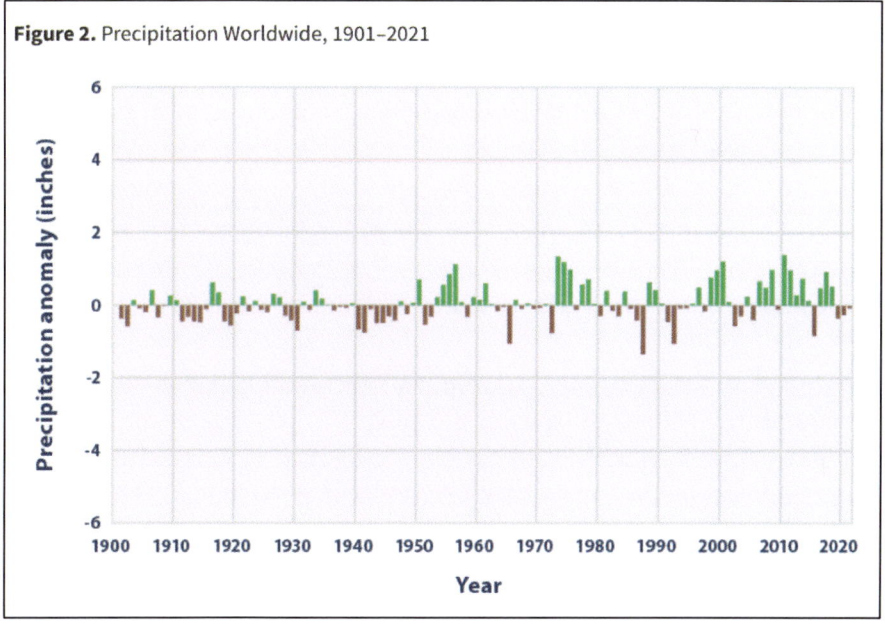

Source: US EPA

Figure 4: Precipitation Worldwide, 1901 - 2021

Hurricanes

The US [EPA Tropical cyclones](#) [Ref 19] web page Figures 1 to 3 examine the frequency, intensity and duration of hurricanes and other tropical storms in the Atlantic Ocean, Caribbean, and Gulf of Mexico. Roughly six to eight hurricanes form in the North Atlantic every year with around two making landfall in the US. There is no overall trend up or down since 1878. However, cyclone intensity has increased over the last 20 years. The EPA include the statement; *"Despite the apparent increases in tropical cyclone activity in recent years, shown in Figures 2 and 3, changes in observation methods over time make it difficult to know whether tropical storm activity has actually shown an increase over time."*

Wildfires

A wildfire is defined as an uncontrolled burn of vegetation, which includes the burning of forests, shrublands and grasslands, savannas, and croplands. Also, wildfires can be caused by human activity — such as arson, unattended fires, or the loss of control of planned burns, and natural causes, such as lightning.

There are many complex issues associated with analysing wildfire trends including assessments of land area burned, carbon emissions, and air pollution. The UK website Our World in Data, Wildfires [Ref 20] web page, see Figure 5, provides a useful data resource for wildfires globally.

In response to the question as to whether there has been an increase in the area burnt by wildfires, the answer provided is *"If you look at statistics from the Global Wildfire Information System shown in the chart here, since the early 2000s, there has been a noticeable decline in the annual extent of land affected by wildfires."* Additional supporting information is provided on the 'Our World in Data' website.

The evidence examined indicates that extreme weather events have increased globally in recent years. it also shows there have been changes in global weather patterns, and across the US in particular with some areas becoming wetter and others drier. Heatwaves, heavy precipitation and flooding have seen increases in frequency, duration and intensity. For hurricanes there is uncertainty as to whether there is any trend up or down, and for wildfires the evidence shows a noticeable decline.

A Guide to the Climate Debate

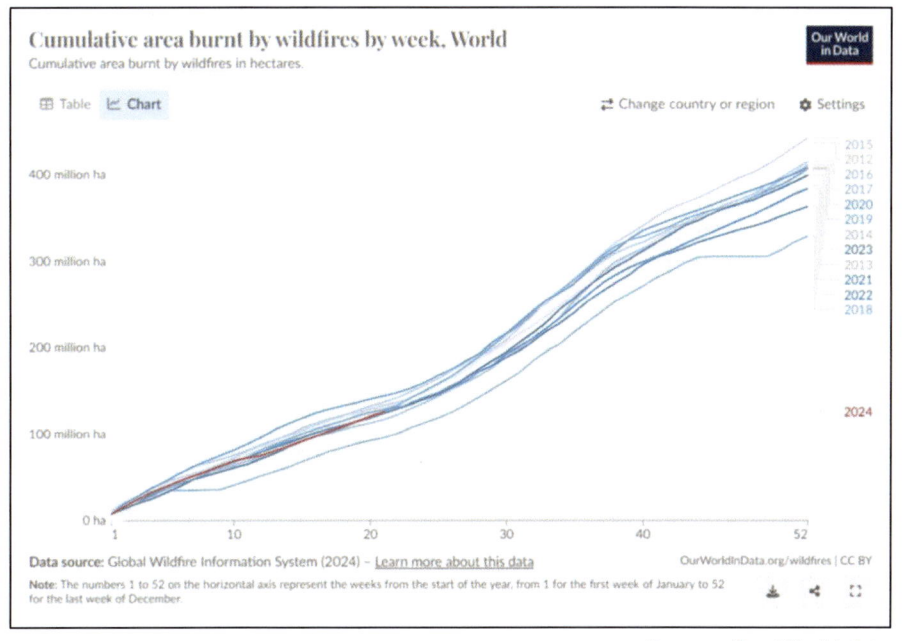

Source: Our World data

Figure 5: Cumulative area burnt by wildfires by week, World

The evidence indicates that extreme weather is increasing globally and will have adverse impacts in some parts of the world if this warming trend continues. Heatwaves, heavy precipitation and flooding are the main risk areas for increased frequency, duration and intensity.

Q3. What is the natural carbon cycle and its relevance?

Carbon is one of the most abundant elements on earth and is essential for life. The earth's carbon exists in many forms and is stored in so-called carbon 'reservoirs' including:

- The atmosphere
- Oceans (also known as the hydrosphere)
- Living things (also known as the biosphere)
- Rocks and soils (also known as the lithosphere)

The Met Office, carbon cycle [Ref 21] web page, see Figure 6, provides an easily understood overview, but for a more detailed account see the Science Direct, Global carbon cycle [Ref 22] web page.

The natural carbon cycle effects the movement of carbon around the earth between these different carbon reservoirs, which in turn helps regulate global temperatures and makes the planet habitable. This is a closed cycle as there is a finite quantity of carbon on the earth, with none being added to or lost from space. The carbon cycle can be considered as being at the heart of life on earth and understanding the carbon cycle is fundamental to understanding climate change.

Carbon in the atmosphere exists mostly in the form of carbon dioxide (CO_2) gas. The (dry) atmosphere is a mixture of gases surrounding the earth consisting of around 78% nitrogen, 21% oxygen and 1% other trace gases including CO_2. By comparison the concentration of CO_2 in the atmosphere is very small, currently around 420ppm, equating to 0.042%. The (wet) atmosphere also contains varying amounts of water vapour, on average about 1%. There are also many tiny particles, solids and liquids called aerosols.

A Guide to the Climate Debate

These aerosols can be dust, spores, pollen, salt, volcanic ash, smoke and pollutants.

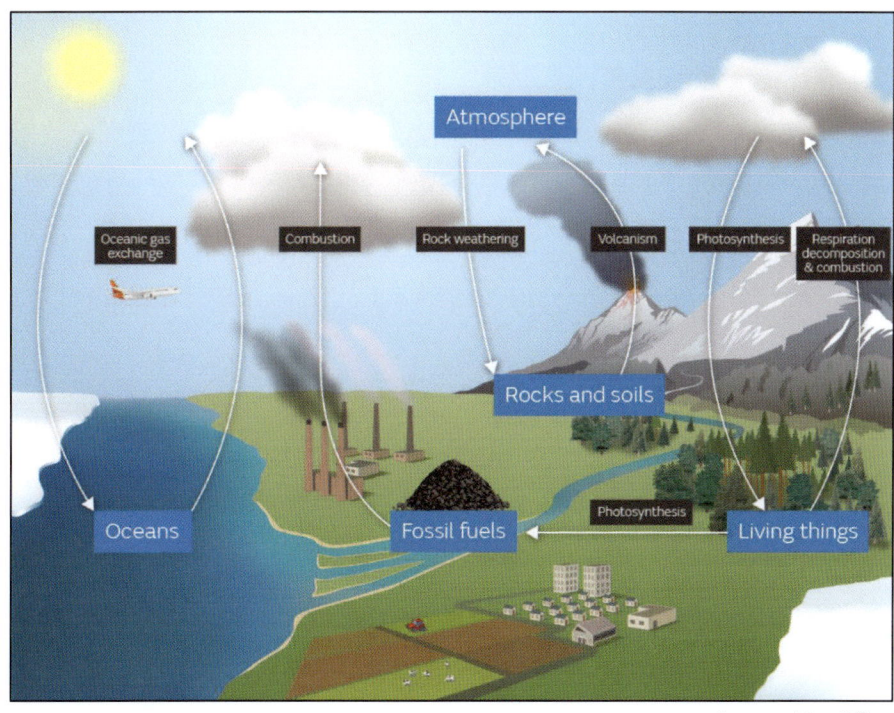

Source: Met Office
Figure 6: Simplified schematic of the global carbon cycle
(Illustrates the main reservoirs of carbon and the processes moving carbon between them)

Contrary to regular and misleading references appearing in the media, atmospheric CO_2 is not poisonous or a pollutant. It is essential to life on earth and is essential for plant growth. (Note; Although CO_2 is not poisonous, a concentration of 40,000 ppm [4.0%] or more is generally considered a dangerous level due to it resulting in oxygen deprivation.)

A Guide to the Climate Debate

CO_2 is also used by the food industry for 'protective atmosphere packaging' to extend the shelf life by excluding the oxygen in normal air. See the Air Products, modified atmosphere packaging web page [Ref 23] for information on the use of CO_2 in food packaging.

Some commercially grown greenhouse crops are grown in an enhanced CO_2 atmosphere to increase growth rates, typically at concentrations of up to 1,000ppm. Refer to the AHDB, CO_2 best practice guide [Ref 24] web page for a horticultural discussion on crop growing supported by increased CO_2. Graph 1 reproduced here as Figure 7 shows that tomato growth increases with increasing CO_2 concentration then levels out at concentrations above 1,000 ppm.

One of the benefits of an increased atmospheric CO_2 concentration combined with the longer growing seasons associated with global warming is that plant growth, and hence crop yields are increased. Known as CO_2 fertilisation, the increased atmospheric CO_2 concentration is already providing increased crop yields due to the increased rate of photosynthesis. Additionally, climate change is allowing new land areas especially in the northern hemisphere to become suitable for growing crops where previously it was too cold.

Fortunately, the natural system of carbon exchanges that recycles this carbon between the various carbon reservoirs has operated for millions of years maintaining a relatively 'goldilocks' climate situation including the atmospheric CO_2 concentration, enabling life to evolve on earth. Even so conditions on earth have varied between very warm and very cold periods compared to what we may consider to be comfortable today.

In the natural carbon cycle CO_2 is added to the atmosphere from the biosphere (from respiration, decomposition and combustion), from volcanoes (from the lithosphere) and from outgassing by the oceans.

In turn CO_2 is removed from the atmosphere by the biosphere (by plant photosynthesis), by rock weathering and by absorption by the oceans. In more recent times the combustion of fossil fuels through human activities has

A Guide to the Climate Debate

added additional CO_2 to the atmosphere effectively moving additional carbon from the lithosphere to the atmosphere.

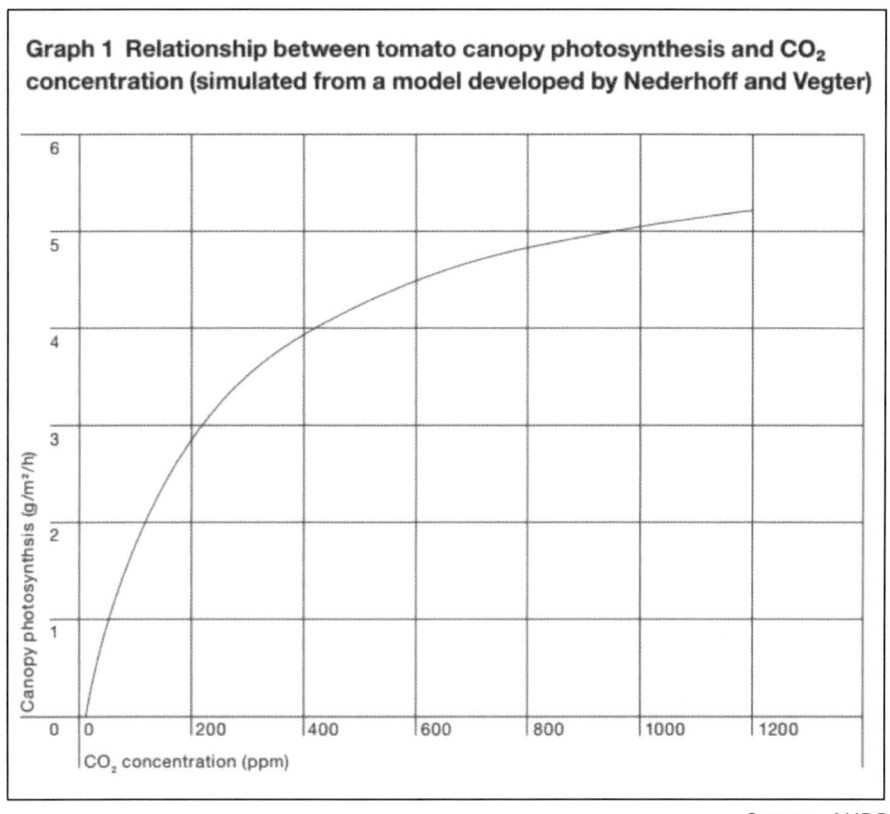

Graph 1 Relationship between tomato canopy photosynthesis and CO_2 concentration (simulated from a model developed by Nederhoff and Vegter)

[Graph showing canopy photosynthesis (g/m²/h) on y-axis from 0 to 6, and CO_2 concentration (ppm) on x-axis from 0 to 1200. Curve rises steeply from origin and levels off around 5-5.5 g/m²/h at higher concentrations.]

Source: AHDB

Figure 7: Relationship between tomato canopy photosynthesis and CO_2 concentration

By far the largest deposits of global carbon are located in the lithosphere but this is considered a relatively inactive component of the natural carbon cycle. These inactive lithosphere deposits are mainly in the form of carbonate minerals and organic compounds including oil, coal and natural gas.

At any one time the bulk of the earth's biologically active carbon is located in the oceans, and biosphere, known as carbon sinks. The distribution of this

A Guide to the Climate Debate

active carbon is around 39,973 x 10^{15} gC (93.15%) in the oceans, 2,190 x 10^{15} gC (5.1%) in the biosphere and 750 x 10^{15} gC (1.75%) in the atmosphere.

Given the enormous difference in amounts of carbon stored in the oceans compared to the atmosphere, just a relatively small percentage loss or gain by the oceans will result in a much larger percentage gain or loss by the atmosphere. Hence the importance of understanding the carbon exchange mechanisms, exchange rates and variabilities of these exchanges.

Carbon is exchanged between the major reservoirs at different rates, there are fast and slow components to the overall carbon cycle. The NASA Earth Observatory, the carbon cycle web page [Ref 25] provides a useful description of the slow and fast cycles. The slow cycles take typically 100 – 200 million years to move carbon between rocks, the ocean and the atmosphere. The slow cycle also returns carbon to the atmosphere through volcanoes.

The fast cycles are measured in months, years and decades and involves the movement of carbon between the biosphere, atmosphere and ocean. Photosynthesis provides the main component of the fast cycle as plants absorb CO_2 and sunlight to create glucose and other sugars. The ebb and flow of the fast carbon cycle is visible in the changing seasons. In the Northern Hemisphere winter, when few land plants are growing and many are decaying, atmospheric CO_2 concentrations climb. During the spring, when plants begin growing again, concentrations drop. It is as if the Earth is breathing.

The ocean plays a key role in our climate system because of its enormous size and relatively rapid exchange of carbon with the atmosphere. The oceans effectively control the atmospheric CO_2 concentration and thereby Earth's climate. Additionally, the ocean takes up solar heat from the atmosphere and moves it around the globe.

The Atlantic Meridional Overturning Circulation (AMOC) which includes the Gulf Stream, moves heat northwards in the Atlantic and this means that Europe is milder than it would otherwise be. These warm surface water and cold deep-water flows have a major influence on the global climate, see Figure 8.

There is evidence that the AMOC varies in strength with a potential risk identified that it may collapse due to global warming. Weakening of the AMOC would reduce the impact of global warming in the northern Atlantic region including the UK. While there appears to be considerable uncertainty regarding these ocean currents, the IPCC has forecast that the AMOC is likely to decline over the 21st Century. Fortunately, there does not appear to be any predictions of imminent AMOC collapse.

The [Met Office, official blog 2 May 2024](#) [Ref 26] provides an update on the AMOC. Additionally, a link to the Marine Climate Change Impacts Partnership (MCCIP) [Ocean circulation report](#) [Ref 27] dated 2023 provides a more comprehensive discussion of the AMOC. The MCCIP reiterates the uncertainties surrounding these ocean flows and the implications of it weakening.

These warm and cold ocean flows will also play a critical role in the carbon cycle affecting the absorption and degassing of CO_2. Cold water tends to absorb CO_2 and warm water tends to release CO_2 through degassing. Given the 53: 1 ratio between the absolute amounts of ocean and atmospheric CO_2, it is evident that a small percentage change in ocean dissolved CO_2 will have a much greater effect on atmospheric CO_2 concentrations.

The [Royal Society briefing 7, the carbon cycle](#) [Ref 28] provides further discussion on these carbon exchanges and the identified uncertainties. The briefing concludes on page 10 with; *"While science has advanced to demonstrate the principles and operation of the carbon cycle, questions remain over the magnitude and timing of impacts of higher atmospheric CO_2 levels, rising temperatures and climate feedbacks on the carbon cycle. Greater understanding of the terrestrial and oceanic carbon sinks and potential sources is therefore a major priority for the decade ahead"*.

A Guide to the Climate Debate

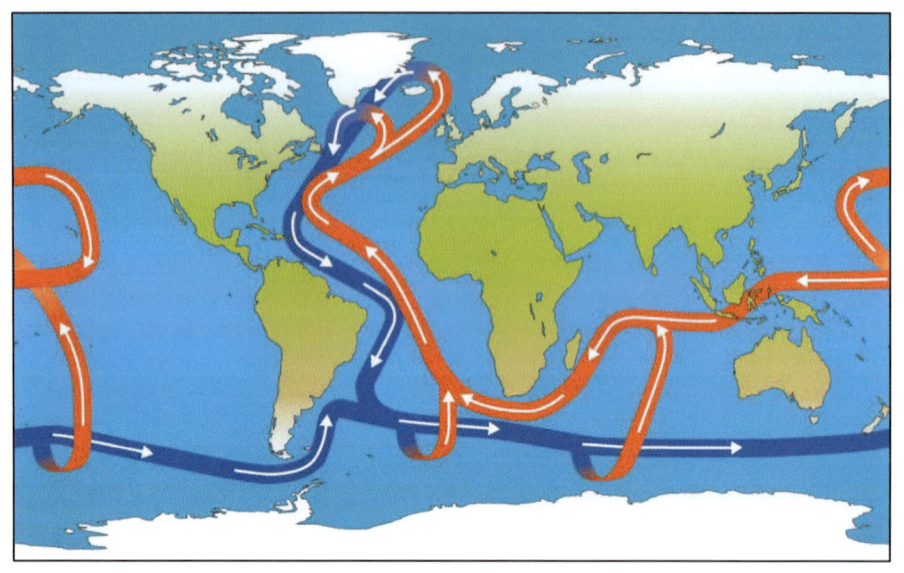

Source: Met Office

Figure 8: Ocean circulations with blue arrows indicating cold deep water and red arrows indicating warm surface water

Despite recent scientific advances in understanding the carbon cycle there are still huge areas of uncertainty as to how it works. These unknowns are acknowledged by scientists and are so significant that they prevent any meaningful climate prediction or forecasts using climate models. Unfortunately, the IPCC fails to properly acknowledge these limitations of climate knowledge in their climate reports and continue to release alarmist messages.

A Guide to the Climate Debate

Q4. Are rising CO_2 levels due to human activity?

Historic climatic data confirms that the level of atmospheric CO_2 has varied over a wide range and has in the distant past exceeded 2,000ppm, although for the last 800,000 years has not exceeded 300ppm, until the 20th century. Each historical peak in CO_2 levels was associated with a rise in global temperature followed by a natural recovery.

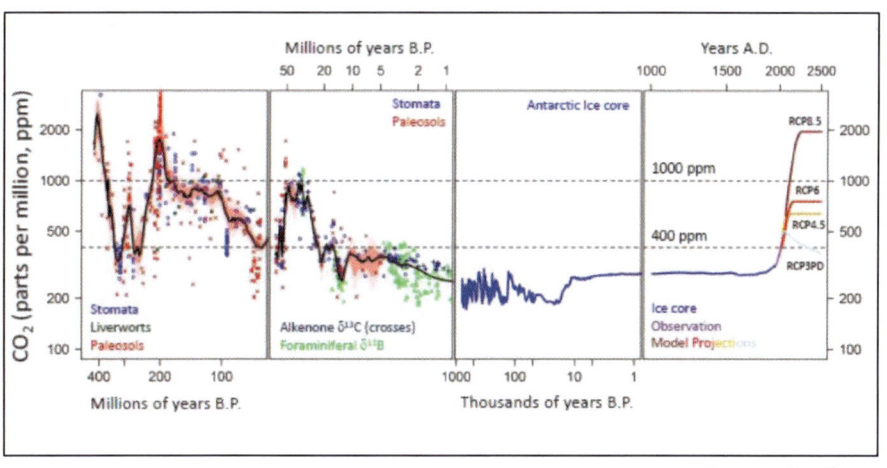

Source: Earth.org

Figure 9: CO_2 levels over the past 500 million years

The Earth.org, A Graphical History of Atmospheric CO_2 Levels Over Time web page [Ref 29] provides a description of the earth's history of CO_2 and climate, see Figures 9 and 10. The NASA, Carbon Dioxide [Ref 30] web page also provides data showing the historical changes in atmospheric CO_2.

A Guide to the Climate Debate

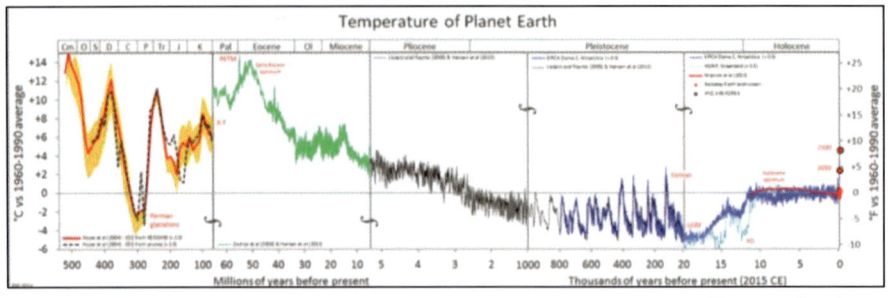

Source: Earth.org

Figure 10: Temperature estimates over the past 500 million years

Pre-industrial levels were around 280pmm and have since risen to 420pmm by 2024. The NASA graphs also show the seasonal variations due to plant growth in the Northern Hemisphere. The NOAA Climate Change, atmospheric carbon dioxide [Ref 31] web page provides further discussion, see Figures 11 and 12.

NOAA claims that each year, human activities release more carbon dioxide into the atmosphere than natural processes can remove, causing the amount of carbon dioxide in the atmosphere to increase. However, this is not quantified.

Note that atmospheric CO_2 concentration is shown to have increased steadily from the 1750 start of the graph at Figure 12 but the rate increased from around 1960. There is also considerable uncertainty as to the overall CO_2 budget (levels, exchanges and emissions) including effects of land-based volcanoes where there are huge measurement margins. The CO_2 emissions from undersea volcanoes is more or less unknown so can be no more than a vague guess. Note, with the oceans covering over 70% of the earth's surface, there is reason to expect that most volcanic emissions are from undersea volcanoes. These uncertainties prevent any meaningful estimate of the actual contribution attributable to human activity.

A Guide to the Climate Debate

CARBON DIOXIDE OVER 800,000 YEARS

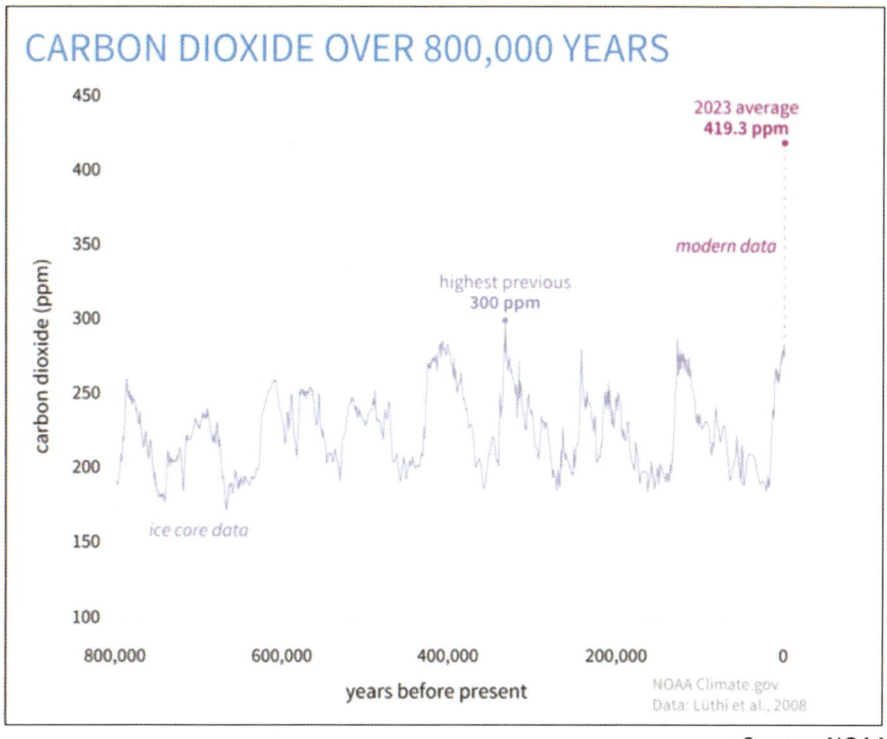

[Graph showing carbon dioxide (ppm) on y-axis from 100 to 450, and years before present on x-axis from 800,000 to 0. Ice core data shows oscillations between roughly 180-280 ppm, with highest previous at 300 ppm. Modern data shows 2023 average of 419.3 ppm. Data: Lüthi et al., 2008, NOAA Climate.gov]

Source: NOAA

Figure 11: Global Carbon Dioxide over 800,000 Years

There appears to be agreement by both Warmists and Sceptics that man-made emissions of greenhouse gasses are mostly the result of the burning of fossil fuels, but also due to deforestation, agriculture and the production of cement. While there appears to be some agreement that human activity is partially responsible for increasing the atmospheric CO_2 concentration, the absolute contributions of natural v manmade sources are unknown.

There is also dispute and uncertainty over the cycle time for atmospheric CO_2 to be exchanged with the ocean and biosphere. While the IPCC claim it can take more than 100 years, there are claims of a much faster exchange of around 5 years. This uncertainty makes a large difference in the proportion of atmospheric CO_2 at any one time attributable to human activity.

A Guide to the Climate Debate

Global atmospheric carbon dioxide compared to annual emissions (1751-2022)

The amount of carbon dioxide in the atmosphere (blue line) has increased along with human emissions (gray line) since the start of the Industrial Revolution in 1750. Emissions rose slowly to about 5 gigatons—one gigaton is a billion metric tons—per year in the mid-20th century before skyrocketing to more than 35 billion tons per year by the end of the century. NOAA Climate.gov graph, adapted from original by Dr. Howard Diamond (NOAA ARL). Atmospheric CO_2 data from NOAA and ETHZ. CO_2 emissions data from Our World in Data and the Global Carbon Project.

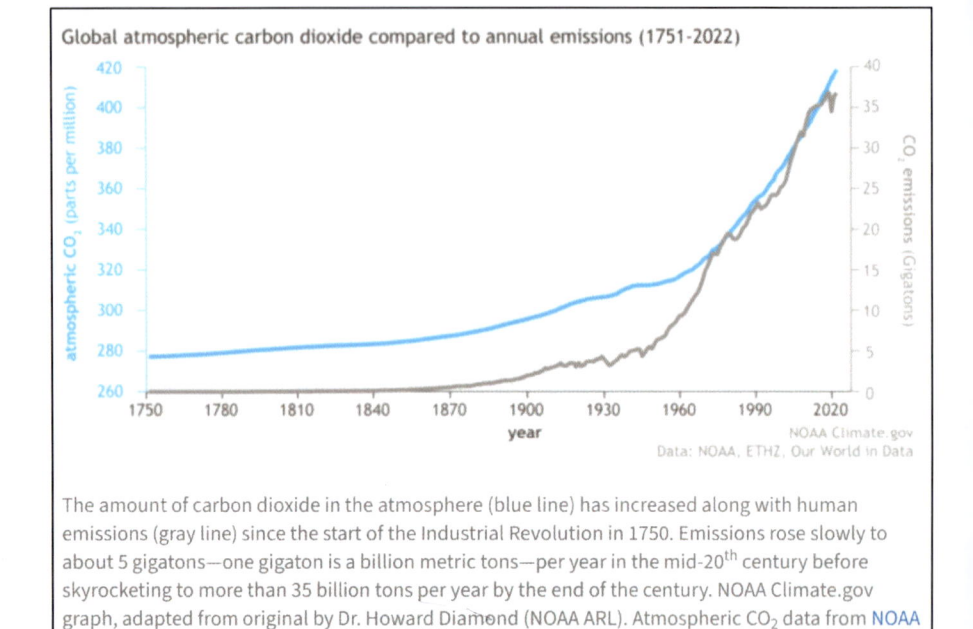

Source: NOAA

Figure 12: Global atmospheric CO_2 compared to annual emissions (1751 – 2022)

The evidence demonstrates that the rising atmospheric CO_2 concentration is caused by a combination of human activity and natural effects. While there is evidence that human activity is partially responsible for increasing the atmospheric CO_2 concentration, there is insufficient evidence to quantify this. The lack of data for volcanos, especially undersea volcanoes makes any meaningful estimates impossible.

A Guide to the Climate Debate

Q5. Does rising atmospheric CO_2 concentration cause global warming?

The theoretical greenhouse effect of the so-called greenhouse gases is defined by the laws of physics and is well documented. The greenhouse gases are so called because they are infrared active, meaning that they absorb solar infrared radiation. This causes heating of the atmosphere so creating the greenhouse effect. The non-greenhouse gases including nitrogen and oxygen are not infrared active so do not contribute to the greenhouse effect. The general principal and underlying physics behind the greenhouse effect appears to be agreed by Warmists and Sceptics.

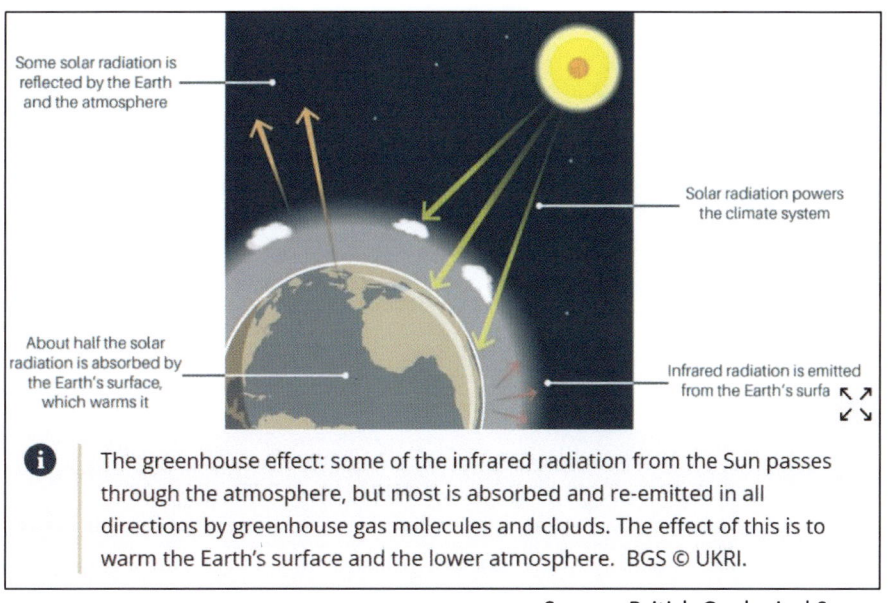

Some solar radiation is reflected by the Earth and the atmosphere

Solar radiation powers the climate system

About half the solar radiation is absorbed by the Earth's surface, which warms it

Infrared radiation is emitted from the Earth's surfa

ⓘ The greenhouse effect: some of the infrared radiation from the Sun passes through the atmosphere, but most is absorbed and re-emitted in all directions by greenhouse gas molecules and clouds. The effect of this is to warm the Earth's surface and the lower atmosphere. BGS © UKRI.

Source: British Geological Survey

Figure 13: The Greenhouse Effect

The British Geological Survey, the greenhouse effect [Ref 32] web page provides a general description. see Figure 13. The main greenhouse gases in order of importance are, water vapour (H_2O) and carbon dioxide (CO_2), followed by methane (CH_4), nitrous oxide (N_2O) and other trace gases. Water vapour comprises around 1% of the atmosphere but is locally highly variable. Local atmospheric concentrations vary from less than 0.01% in extremely cold regions to around 3% by mass in saturated air at about 32°C.

> Water vapour is the most abundant and most important greenhouse gas although the IPCC does not consider it a greenhouse gas, instead they refer to it as a 'forcing agent'. This appears to be an IPCC marketing tactic in order to maintain the focus on CO_2 as the global warming villain.

However, science tells us that water vapour is a greenhouse gas and it is the most dominant greenhouse gas by a significant margin. Evidence shows the relative indicative greenhouse effect with a clear sky (Note that cloud cover will have a moderating effect on warming through the reflection of solar radiation back out to space) would be: water vapour 67%, CO_2 24%, others 9%.

The atmospheric energy flows are highly complex with the 2009 paper by Trenberth et al. titled Earth's Global Energy Budget [Ref 33] providing a more detailed discussion of the global atmospheric energy budget. Trenberth's Fig 1 reproduced here as Figure 14 provides a diagrammatic representation of the various global energy flows. It is notable that the atmosphere is heated mostly from below, by water vapour absorption of outgoing infra-red radiation, by latent heat transfer by evapo-transpiration and condensation and direct warming of the air. This heat balance diagram also shows the energy incoming (341) is balanced by the outgoing (102 + 239 =341).

Andrew Montford in his book **'The Hockey Stick Illusion'** [Ref E] discusses at page 391 the effects of cloud cover on global warming, referencing a 2006 paper by Sandrine Bony titled, How well do we understand and evaluate climate change feedback processes? [Ref 34]. There is evidence that a warming climate

A Guide to the Climate Debate

generates a substantial increase in cloud cover and that this is known to have a strong negative, or cooling effect on the climate. Montford quotes Bony:

Global Energy Flows W m^{-2}

102 Reflected Solar Radiation 101.9 W m^{-2}	341 Incoming Solar Radiation 341.3 W m^{-2}	239 Outgoing Longwave Radiation 238.5 W m^{-2}

Reflected by Clouds and Atmosphere 79

Emitted by Atmosphere 169

40 Atmospheric Window

Absorbed by Atmosphere 78

Greenhouse Gases

Latent 80 Heat

17

Reflected by Surface 23

40

356

333 Back Radiation

161 Absorbed by Surface

17 Thermals

80 Evapo-transpiration

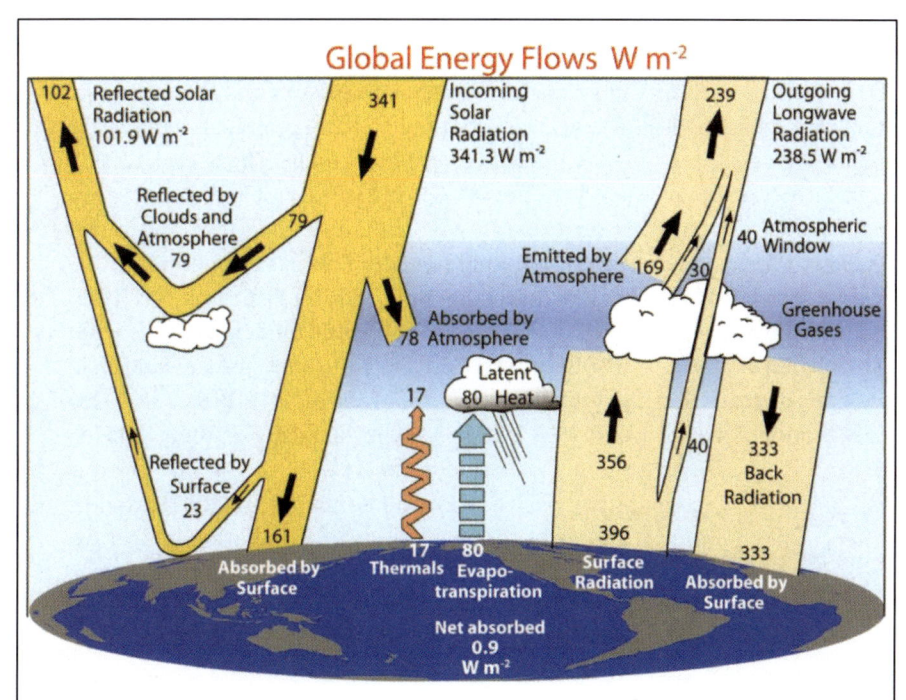

396 Surface Radiation

333 Absorbed by Surface

Net absorbed 0.9 W m^{-2}

FIG. 1. The global annual mean Earth's energy budget for the Mar 2000 to May 2004 period (W m^{-2}). The broad arrows indicate the schematic flow of energy in proportion to their importance.

Source: Trenberth *et al.*

Figure 14: Global Energy Flows

"Boundary layer clouds have a strongly negative [feedback effect]..... and cover a very large fraction of the area of the Tropics.... Understanding how they may change in a perturbed climate therefore constitutes a vital part of the cloud feedback problem. Unfortunately, our understanding of the physical processes that control boundary layer clouds and their radiative properties is currently very limited"

> Bony also states *"Cloud feedbacks have long been identified as the largest internal source of uncertainty in climate change predictions.....".*

On page 393 Montford again touches on the way clouds affect the climate, and how it was reported in the IPCC Fourth Assessment Report (FAR). Noting again that there is evidence that a warming climate generates a substantial increase in cloud cover and that this is known to have a strong negative, or cooling effect on the climate. However, the FAR downplayed the significance of this negative effect, focussing only on the positive feedback effects. These positive forcing effects are a key element supporting the Warmist global warming hypothesis as the warming effects from manmade greenhouse gas emissions are insufficient on their own and it seems it can only be shown to be potentially catastrophic if only the positive forcings are considered.

Despite the water vapour greenhouse effect dominance, very little mention of its contribution to the overall greenhouse effect appears to made in the literature and websites reviewed. Instead, the IPCC and Warmists focus primarily on CO_2, which is in effect a relatively minor greenhouse gas.

There is also evidence that the greenhouse effect from CO_2 is not a linear function. In fact, it is known to be a logarithmic function. This means that each incremental amount of CO_2 exerts a logarithmically smaller additional warming effect. In practice the global greenhouse effect is much more complex and unpredictable than the theoretical science can explain.

> As a result of this logarithmic relationship, any further increases in atmospheric CO_2 from current levels are expected to result in only a few more tenths of a degree of additional warming before capping out. The alarmist predictions of ever increasing warming due to a rising atmospheric CO_2 concentration is therefore vastly over exaggerated.

Warmists are confident in their claim that the main cause of global warming is the rising concentration of greenhouse gases in the atmosphere, mainly CO_2

A Guide to the Climate Debate

and that this is due to emissions caused by human activities. Warmists support their argument with predictions from sophisticated computer models. However, these models will only be as good as the data fed into them, data that we know contains large amounts of uncertainties. Warmists do acknowledge that there are natural causes accounting for part of the current global warming but do not know to what extent.

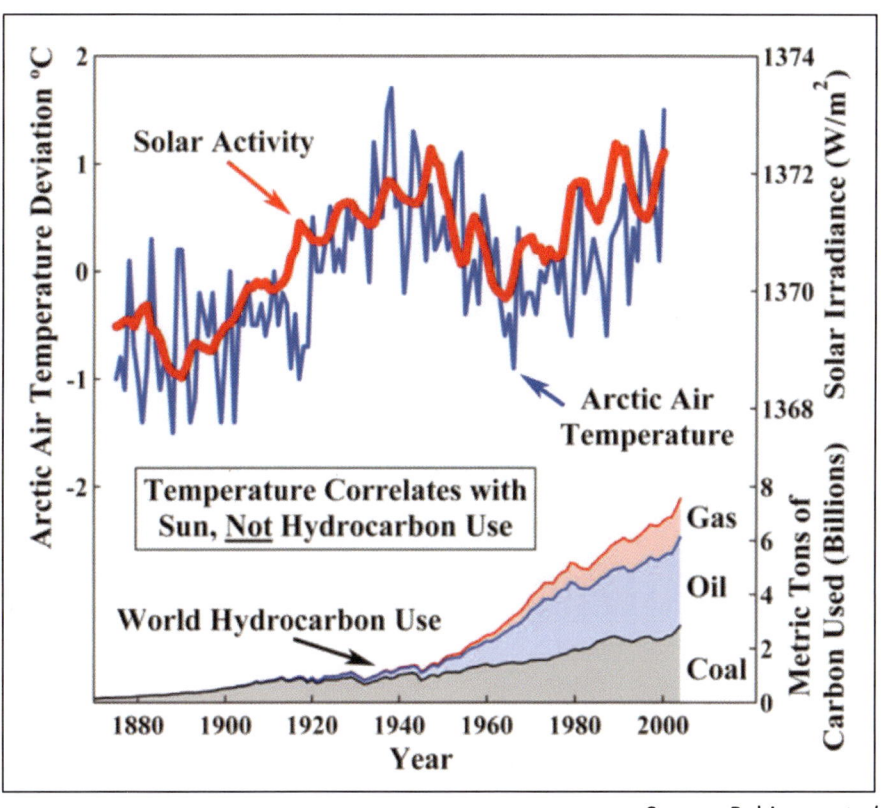

Source: Robinson *et al*.

Figure 15: Arctic surface air temperature compared with total solar irradiance as measured by sunspot cycle amplitude, sunspot cycle length, solar equatorial rotation rate, fraction of penumbral spots, and decay rate of 11-year sunspot cycle. Solar irradiance correlates well with Arctic temperature, while hydrocarbon use does not correlate.

A Guide to the Climate Debate

Sceptics claim that the rising atmospheric CO_2 concentration has a much smaller input to the overall greenhouse effect and that rising temperatures are driven mainly by solar activity. Also, that the rising atmospheric CO_2 concentration is largely a response to the rising temperatures effecting the natural carbon cycle exchanges, principally outgassing of CO_2 from the ocean as it warms. Sceptics base their argument partially on the historic data.

Referring to the paper Environmental Effects of Increased Atmospheric Carbon Dioxide [Ref 35] by Robinson *et al*. 2007, Figure 3 (reproduced here as Figure 15) makes the claim that global temperature correlates with solar activity, but does not correlate with the world's hydrocarbon use. The author's quote;

"Correlation does not prove causality, but non-correlation proves non-causality", and *"The experimental data do not prove that solar activity is the only phenomenon responsible for substantial Earth temperature fluctuations, but they do show that human hydrocarbon use is not among those phenomena"*. Note that Figure 3 (Figure 15 here) covers a time period from 1880 to 2000.

In contrast, a 'Warmist' view is provided by the Royal Society What role has the Sun played in climate change in recent decades? [Ref 36] web page and provides a very different explanation. Figure 2 (reproduced here as Figure 16) and plotting the sun's energy and global temperature claims to show no correlation. However, the time period of the plot is limited to the period 1978 to 2018. If we compare the Royal Society curves with the same 1978 to 2018 section from the Robinson paper Figure 3 (1880 to 2020), we find the actual data for that 1978 to 2018 time period are very similar. This begs the question as to why the Royal Society restricted the data to the period from 1978 to 2018?

There is also a correlation between the Robinson Figure 3 (Figure 15) and the Central England Temperature (CET) data, Met Office, The Central England Temperature series [Ref 37] web page adding further credibility to the Sceptic interpretation.

A Guide to the Climate Debate

This comparison demonstrates that simply through adjustment of the analysis time period we can claim very differing outcomes when using the same data. Comparable to say comparing the analysis of a single jigsaw piece against the whole jigsaw picture, demonstrating how data can be manipulated to provide the desired outcome.

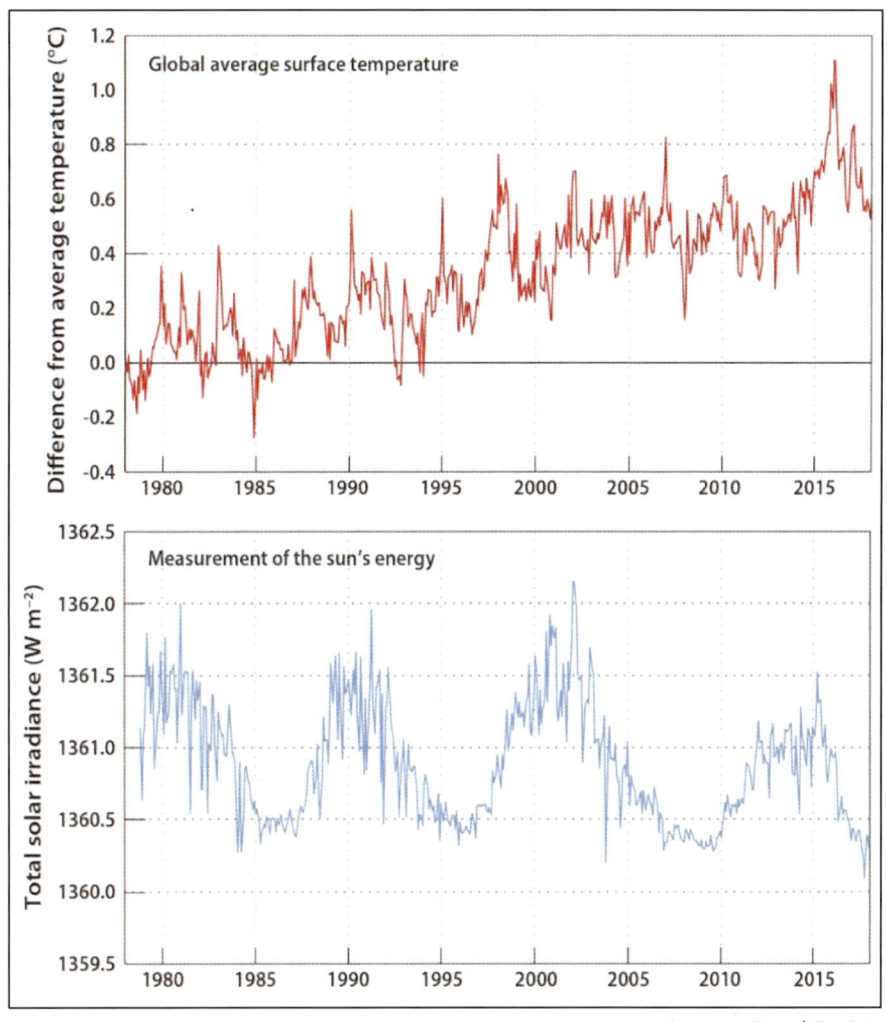

Source: Royal Society

Figure 16: Global average temperature and measurement of the sun's energy (1978 – 2018)

A Guide to the Climate Debate

On the assumption that the underlying data is correct, then the Robinson *et al.* conclusion would appear to be more plausible and adds credibility to the theory that the cycles of solar activity play the major role in driving climate change, and not atmospheric CO_2.

There is a proven correlation between solar activity and global temperature, but no correlation between the global use of hydrocarbon and global temperature. Whereas correlation does not prove causality, non-correlation proves non-causality. **This evidence is inconsistent with the IPCC climate change hypothesis.**

Q6. How does the sun influence the climate?

The earth's surface where we live has two main sources of heat energy. A relatively small amount of heat comes from within the planet, from the molten core and decay of radioactive elements. However, the dominant heat source is the sun and this heat energy subsequently redistributed by radiation and convection is what drives the earth's climate system. This total solar radiation of light and heat known as the 'Total Solar Irradiance' (TSI) is not constant, but varies through two mechanisms;

- First, there are internal stellar processes that affect the radiant energy emitted by the sun, i.e. solar activity.
- Second, changes in the earth's orbit around the sun over tens and hundreds of thousands of years that directly affect the amount of solar energy hitting the earth.

Solar activity is characterised by the 11-year sunspot cycle (but can be 9 to 13 years), with increasing numbers of sun spots resulting in an increase in TSI. There has been a small long-term trend for TSI to increase.

Changes in the earth's elliptical annual orbit around the sun are known as the Milankovitch Cycles. These cycles affect the eccentricity, tilt and precession of the earth's orbit and hence distance from the sun and this affects the amount of energy hitting the earth. These solar cycles correlate with changes in the climate identified over the last 400,000 years from ice cores.

The Imperial College London briefing paper No. 5 dated 2011 titled: Solar Influences on Climate [Ref 38] by Professor Joanna Haig provides a detailed overview of this subject and Figure 2, reproduced here as Figure 17 illustrates the Milankovitch cycles. What this paper has highlighted are the gaps in the knowledge regarding how the sun influences the earth's climate.

A Guide to the Climate Debate

While clearly a 'Warmist' authored paper that carefully avoids any challenge to the Global Warming hypothesis, it is quite critical on page 17 of the general circulation models (GCMs) used by the IPCC in their assessment reports. It ends with the typical scientist's understatement,

"Thus, a number of improvements in climate models are likely to become necessary in order for them to provide a more faithful representation of solar impacts".

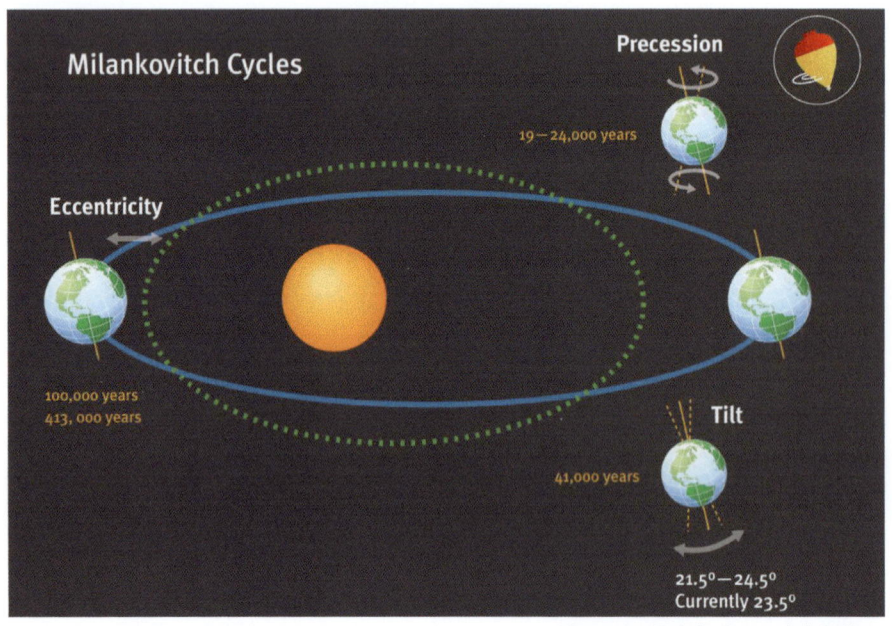

Source: Haigh, Imperial College

Figure 17: The earth's orbit around the Sun, showing the key parameters eccentricity, tilt and precession

Uncertainties identified in this Haig paper include;

- Cloud cover is recognised as having a large effect by reflecting solar radiation and so tending to cool the earth's surface. Also, a recognition that atmospheric ionisation due to cosmic rays could

influence the amount of cloud cover. Large uncertainties surround the prediction of cloud cover and its effect on the climate.
- The effects of galactic cosmic rays (GCRs) originating outside the solar system and the uncertainties of the effects. There is an inverse relationship between solar activity and the incidence of GCRs.
- Uncertainties with the calibration of instruments measuring the absolute value of TSI. With the measurement uncertainties greater than the variations being measured, this is regarded as a serious problem underlying solar-climate research.
- Uncertainties surrounding the variability of the solar wavelength spectrum and hence how it heats the atmosphere through the greenhouse effect.

In the documentary Climate the Movie [Ref 1] evidence was presented showing a correlation between galactic cosmic radiation (GCRs), solar TIR and global temperature variations. What the Imperial College paper confirms is that GCRs do play a significant role in the global climate variations and should not be excluded from consideration as is current IPCC practice.

The paper IPCC Underestimates the Sun's Role in Climate Change [Ref 39] by van Geel & Ziegler, 2013, provides further evidence to challenge the IPCC view on the solar effect. In its conclusion, the van Geel paper states;

"There is no doubt that during the past 40 years atmospheric CO_2 concentrations increased, partly owing to fossil fuel consumption. During the second half of the 20th century solar activity has, however, also reached a very high level, with TSI increasing considerably more than what the GCMs allow for. Indeed, the long-lived Grand Modern Solar Maximum that had commenced in 1924, culminated in the second half of the 20th century and ended in 2008 involved TSI fluctuations of up to 3 Wm-2. Moreover, the GCMs conveniently overlook the fact that the 20th century temperature increase involved, apart from an important TSI increase, also an increase in solar UV radiation and a decrease in the GCR flux, as well as their fluctuation during the different PDO phases. In the 5th Assessment report of the IPCC, models assuming natural climate forcing only, indicate no temperature increase

during the 20th century. By contrast, Holocene paleo-climatologic data indicate that the Earth's climate is very sensitive to even small fluctuations in solar radiation. In view of the observed TSI increase during the 20th century, it is therefore highly unlikely that temperatures would have remained stable during this time, provided greenhouse gas emissions had remained stable. Indeed, the observed cyclical temperature rise of the 20th century can be explained by the buildup of the modern Grand Solar Maximum that culminated in the 1990s and early 2000s. This suggests that the climate models of the IPCC, as presented in its 5th Assessment Report (high sensitivity to greenhouse gases, low sensitivity to TSI variations), must be erroneous due to neglect of the observed TSI variation."

Also;
"The IPCC dismisses the important role of the Sun in natural climate change as evidenced by the climate history and blames mankind for climate change without presenting convincing physical evidence that increasing atmospheric CO_2 concentrations are indeed the cause of global warming, a highly controversial subject".

The evidence reviewed regarding the effects of solar radiation highlights the uncertainties with virtually every aspect of the global heat cycle. These uncertainties include, effects of cloud cover, effects of galactic cosmic rays, uncertainties with the variability of the solar wavelength spectrum, and uncertainties with the actual measurements of solar TSI. These uncertainties prevent any meaningful prediction or forecasting of climate changes with IPCC GCMs. **This evidence is inconsistent with the IPCC climate change hypothesis.**

Q7. Does rising atmospheric CO_2 concentration lead or lag warming?

This is a crucial question when attempting to validate the IPCC climate change hypothesis. Warmists claim that rising CO_2 levels are driving the rise in global temperature due to the greenhouse effect. This scenario requires the CO_2 concentration to increase first and then the temperature to rise in response. Also, by implication, if we can lower the atmospheric CO_2 concentration by lowering or eliminating man-made CO_2 emissions, then we will reverse or stop global warming. This claim lies at the heart of the IPCC climate change hypothesis, underpinning the Paris Agreement and subsequent Net Zero Policies.

Sceptics claim that global temperature is driven by solar activity and that the rising temperature subsequently drives up the atmospheric CO_2 concentration, largely due to outgassing by the ocean. For this scenario the temperature must increase first and then the CO_2 concentration to increase in response. Also, by implication, eliminating man-made CO_2 emissions will have little or no effect on global warming.

In the documentary Climate the Movie [Ref 1] it is claimed that ice core data provides historical evidence that rising temperature was found to occur before CO_2 concentrations increased, the time lag being typically measured in hundreds of years. The conclusion being that this proves that rising CO_2 is a consequence of rising temperature, and not rising temperature being a consequence of rising CO_2 and hence disproves the IPCC climate change hypothesis. The movie also claims that rising temperature is caused by a variation in solar radiation. The only way to prove this claim and hence challenge the IPCC climate change hypothesis, is for it to be tested with actual (empirical) climate data covering a long enough period. To gain a better understanding and to determine cause and effect it is necessary to examine the historical climate records.

A Guide to the Climate Debate

Although direct and continuous measurement of global weather data only goes back around 150 years, there are now proxies extending the data into the distant historical past. The most distant period of time for which there are estimates of CO_2 levels and temperatures are 500 million years, see Figures 9 & 10. However, more robust evidence obtained from ice cores going back around 800,000 years provides high resolution records from analysis of the air bubbles trapped in the ice.

Additional information on the ice core data is provided on the British Antarctic Survey Ice cores and climate change [Ref 40] web page. Figure 3 shows the variation of CO_2 and temperature over the 800,000-year period and it states that temperature and CO_2 are intimately linked, but omits to say how they are linked. The British Antarctic Survey page also claims that CO_2 starts increasing before temperature with Figure 4 providing the basis of this claim. However, on examining the graph, this claim appears to be questionable. The graph at Figure 4 covers only the last 23,000 years out of the 800,000 years of the available data. Even at this relatively short and hence stretched time axis it is not clear that during the rise in temperature and CO_2 from around 17,000 years ago to around 10,000 years ago that CO_2 does actually lead temperature as claimed.

Ice core data analysis has also been found on the claimed independent Climate Data Information website titled, Climate Data Information, Ice cores [Ref 41]. This website provides further explanation of the ice core method and analysis of the data covering the last 800,000 years. Figure 2 shows temperature and Milankovitch cycles and includes the statement; *"Figure 2 shows the temperature and radiation data as calculated from the Milankovitch cycles. As can be seen, there is good agreement between the timing of orbital cycles and the temperature changes. It is also noticeable that there is a tendency for times of higher radiation to be associated with higher temperature but the correspondence is far from perfect"*. This statement would appear to agree with the claim made by Robinson, Robinson & Soon in Environmental Effects of Increased Atmospheric Carbon Dioxide [Ref 35] that global temperature fluctuations correlate with changes in solar radiation.

A Guide to the Climate Debate

Climate Data Information, Ice cores [Ref 41] Figure 3, reproduced here as Figure 18 shows temperature and CO_2 over the same 800,000-year period and includes the statement; *"Figure 3 shows temperature and CO_2. As can be seen, there is close correspondence in the timing and relative magnitude of the two variables. Although not clear from this chart there is general agreement that temperature changes precede CO_2 changes during the rising phase and CO_2 lags temperature by a larger amount during the falling phase. Figure 4 and Figure 5 show similar relationships for the Vostok ice core (note: EPICA and Vostok are both in the Antarctic)".*

Note that the Climate Data Information web page carefully steers an independent path avoiding any Warmist – Sceptic bias by confining its report to the data analysis only. It does not comment on the implications of the CO_2 and temperature changes. These two statements on the Climate Data Information web site would appear to confirm the claim in **'Climate the movie'** [Ref 1] that global temperature fluctuations are driven by solar radiation and that atmospheric CO_2 concentration is responding and following temperature. Additionally, their Figure 3, reproduced here as Figure 18, peaks and troughs would appear to correspond to the British Antarctic Survey analysis [Ref 40] Figures 2 and 3 graphs, providing some confidence that both are using the same data.

It seems inconceivable that had clear evidence been obtained from the ice cores of rising CO_2 preceding rising temperature, that the IPCC would not have held back from announcing a ground breaking discovery to prove the climate change hypothesis. This has never happened and poses the question as to why, with what appears to be **conclusive proof that instead, temperature leads CO_2, so disproving the IPCC climate change hypothesis,** has this information not been publicised?

An article in New Scientist titled Climate myths: ice-cores show CO_2 increases lag behind temperature rises disproving the link to global warming [Ref 42] dated 16 May 2007 by Michael Le Page and Catherine Brahic attempts some explanation of this analysis, again conflicting with the IPCC climate change hypothesis.

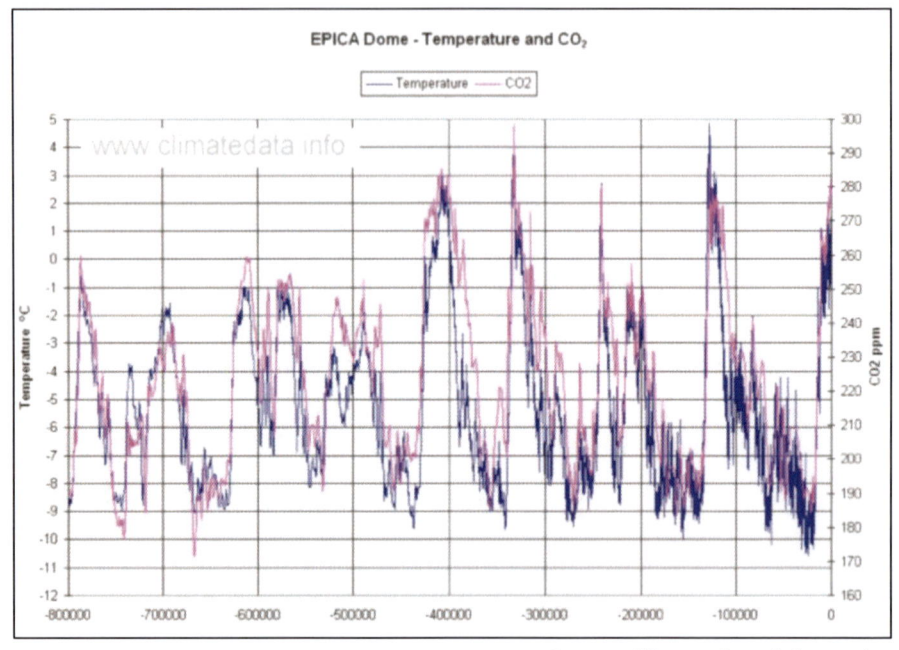

[Chart: EPICA Dome - Temperature and CO₂, showing Temperature (°C) and CO2 (ppm) over time from -800000 to 0. Source: www.climatedata.info]

Source: Climate Data Information

Figure 18: Temperature and CO$_2$.
("As can be seen, there is close correspondence in the timing and relative magnitude of the two variables. Although not clear from this chart there is general agreement that temperature changes precede CO$_2$ changes during the rising phase and CO$_2$ lags temperature by a larger amount during the falling phase.")

While this is clearly a 'Warmist' article with a title seemingly aimed at discrediting Sceptic claims, it provides some helpful clarifications. It is suspected that these authors are closet Sceptics surviving in a Warmist scientific world. The article also speculates on other possible causes of this amplification effect and suggests there must be some unknown limiting factors that prevent a runaway greenhouse effect.

A Guide to the Climate Debate

> There is an acknowledgement in the New Scientist article that CO_2 usually started to rise only after temperatures had begun to climb so confirming the 'Climate Data Information' observations and the claim from 'Climate the Movie'. The New Scientist article also acknowledges that the Milankovitch solar cycles seem to have triggered the beginning and end of many ice ages, but that this alone cannot explain the full extent of the temperature changes. It suggests that CO_2 helped to amplify the changes.

The conclusions from all this evidence are that the global warming and cooling cycles are initiated by the Milankovitch cycles and hence the cycles in solar radiation. However, it is claimed that the cycles of solar radiation through the Milankovitch cycle effect would be insufficient on its own to power the warming phase. Rising CO_2 levels appear to be a consequence of the warming (mainly through degassing by the ocean) and must have provided some amplification, or forcing through the greenhouse effect. However, this forcing effect should, according to greenhouse theory have created a positive feedback loop increasing the rate of temperature rise and hence leading to thermal runaway.

This thermal runaway has never happened and what has prevented thermal runaway and then forced the warming phase into a cooling phase as the Milankovitch cycle turned is unknown as the New Scientist article acknowledges.

> Importantly, the cooling phases initiated while the CO_2 concentration was at a peak high value, and likewise the warming phases were initiated when CO_2 concentration was at a minimum value, **again conflicting with the IPCC climate change hypothesis.** Although there is plenty of speculation as to why this could occur, it is another important uncertainty in the understanding of how the climate system works. It is most likely that atmospheric CO_2 concentrations do play a role in reinforcing both the warming and cooling cycles but in a minor way and in a way that is not understood.

A Guide to the Climate Debate

Q8. What about sea level rise and ocean acidification?

Warmists claim that sea levels are rising at unprecedented rates and as a consequence of man-made global warming. The threat of sea level rise has been accompanied by various alarmist warnings of flooding of low-lying islands and populated coastal areas around the world. These claims of catastrophic sea level rise and ocean acidification need to be examined and the facts put into the correct context.

Sea level rise

Before we can understand the risk from rising sea level we need to clarify the definitions of sea level. The NOAA web page Tracking sea level rise and fall [Ref 43] provides a description;

- 'Global sea level' (or Global Mean Sea Level) is the average height of the ocean around the world.
- 'Mean sea level', also known as the tidal datum is the average of the hourly water levels observed at a local tide station over 19 years.
- 'Local sea level' is the height of the water measured on the coast relative to a specific point on land.

When we assess global sea level rise, we need to consider first the historic records. The NOAA, Climate Change: Global Sea Level [Ref 44] web page provides a helpful graph, reproduced here as Figure 19 showing the change in sea level since 1880. Over this period sea level has varied up and down but shows a consistent overall rise, having risen around 210 to 240mm during the period. Also, the rate of increase recorded has increased from around 1.4mm per year

A Guide to the Climate Debate

during the 20th century to around 3mm per year over the last few decades. This historic data indicates that global sea level was rising well before the acknowledged increase in carbon emissions due to human activities.

The NASA Sea Level Change Understanding Sea Level, Key Indicators [Ref 45] website (Global Mean Sea Level page) shows the sea level has risen 103.3mm since 1993, an average of 3.4mm per year. Based on this historic data, there is a reasonable expectation for a similar rate of increase continuing over the next few decades. Further information is provided on the IPCC, Sea Level Rise [Ref 46] web page where the paper by Warrick and Oerlemans titled Sea Level Rise [Ref 47] can be downloaded. Although this paper is undated (likely pre-2000) it is thought to be relevant as this is a current IPCC webpage. In the executive summary the paper states;

- *"The average rate of rise over the last 100 years has been 1.0 - 2.0 mm yr. "*
- *"There is no firm evidence of accelerations in sea level rise during this century (although there is some evidence that sea level rose faster in this century compared to the previous two centuries)"*
- *"As to the possible causes and their specific contributions to past sea level rise, the uncertainties are very large, particularly for Antarctica. However in general it appears that the observed rise can be explained by thermal expansion of the oceans and by the increased melting of mountain glaciers and the margin of the Greenland ice sheet. From present data it is impossible to judge whether the Antarctic ice sheet as a whole is currently out of balance and is contributing, either positively or negatively, to changes in sea level."*

The evidence reviewed appears to agree that the global sea level is rising although there are some uncertainties as to the accuracy of the data. Historically sea level was measured using tide gauges located around the world with some records going back around 300 years. From these historic records, sea level was reconstructed with corrections applied to allow for land sinking and uplift or for other effects that could raise or depress the sea level locally. The world's ocean is not level as is the water in a small pond or bath tub. It is a mosaic of peaks and troughs resulting from variations in gravity, the earth's rotation, also the effects of temperature, wind and ocean currents.

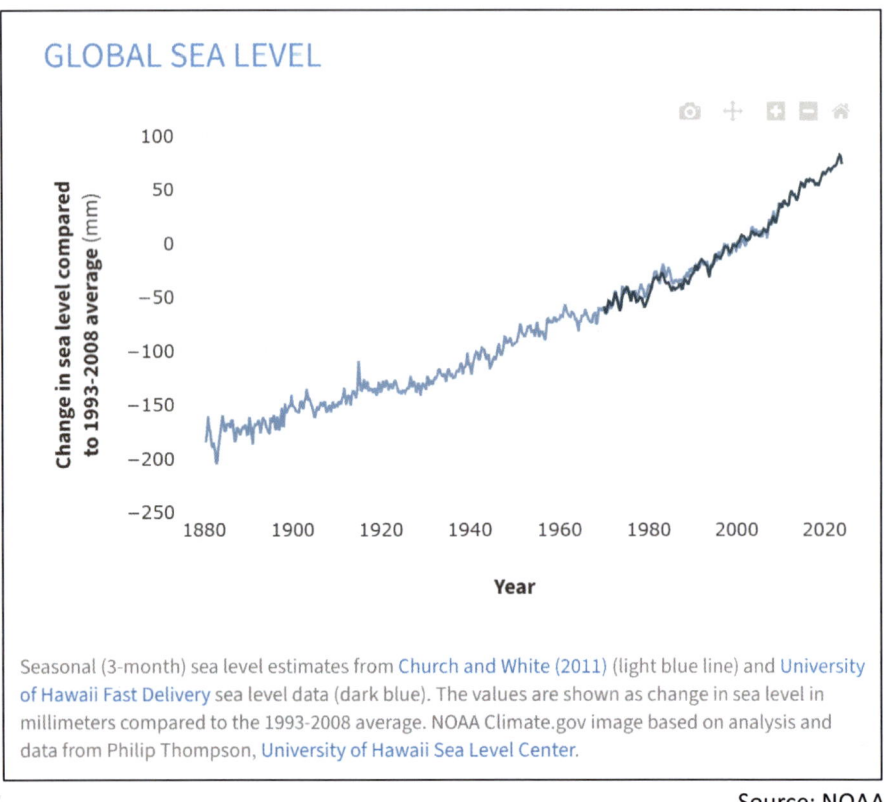

GLOBAL SEA LEVEL

Seasonal (3-month) sea level estimates from Church and White (2011) (light blue line) and University of Hawaii Fast Delivery sea level data (dark blue). The values are shown as change in sea level in millimeters compared to the 1993-2008 average. NOAA Climate.gov image based on analysis and data from Philip Thompson, University of Hawaii Sea Level Center.

Source: NOAA

Figure 19: Global Sea Level changes

Estimates from these historic records for the 20th century show an average sea level rise of 1.6mm per year. Since 1992 measurements of sea level have been made by satellites with corrections applied and this new method of measurement has shown an annual rise of around 3.2mm per year, two times the annual rise from conventional measurements. The reasons for this significant difference are unknown but adds to the uncertainty when evaluating seal level rise. The above mentioned paper by Warrick *et al*. [Ref 47] examines the evidence for acceleration in the rate of sea level rise at section. 9.3.3 where they state;

"From examinations of both composite regional and global curves and individual tide gauge records, there is no convincing evidence of an acceleration in global sea level rise during the twentieth century. For longer periods, however, there is weak evidence for an acceleration over the last 2-3 centuries."

The evidence indicates that the main causes of long-term sea-level rise are:

- Thermal expansion of the ocean due to its warming.
- Melting of land-based ice from mountain glaciers and ice sheets. It should be noted that melting sea ice does not contribute to sea level rise as it is already floating on and is already part of the ocean.

Thermal expansion of the ocean

The ocean covers over 70% of the earth's surface and is over 3km deep over much of its area. Water is one thousand time denser than air and has far more mass than the atmosphere. As a result, the ocean has around 3,300 times greater heat capacity than the atmosphere. Due to this thermal disparity, it is not possible for the atmosphere to exert any significant heating effect on the ocean as is sometime claimed by Warmists. Instead, it is the ocean that controls the warmth of the lower atmosphere via;

- Direct contact
- Infrared radiation from the ocean surface
- Removal of latent heat by evaporation (so cooling the water)

The earth's climate control system revolves around the movement of heat energy received from the sun and in maintaining a heat balance as discussed at key questions 5 and 6. All heat absorbed by the earth from the sun is eventually lost to space to maintain this balance. The ocean absorbs heat from the sun and the atmosphere and so is warmed, but is also subject to

A Guide to the Climate Debate

increased evaporation which exerts a cooling effect through the removal of latent heat so limiting the warming effect. It is therefore the ocean that exerts the largest effect in determining how heat received from the sun is absorbed and recirculated around the globe via ocean currents.

A key factor in this redistribution and climate control are the inherent time lags for the atmosphere and the ocean. The time constant for the atmosphere for a molecule of CO_2 to be recirculated is claimed to in the order of one year. The similar time constant for the ocean is in the order of one thousand years. This results in major time lags in the earth's climate system.

> There are suggestions from some scientists that the rise in the atmospheric CO_2 concentration during the 20th century represents ocean outgassing caused by the Mediaeval Warm Period after a 1,000-year time lag.

While there is some agreement that a warming ocean is a major contributor to sea level rise, estimating the ocean warming contribution is highly complex and subject to considerable uncertainties. There is no evidence that the rate of sea level rise is increasing significantly or that human activities are having any measurable effect.

Glaciers and ice sheets

The majority of valley glaciers have been retreating (melting) over the last 100 years with a trend identified of worldwide glacier retreat since the Little Ice Age. The evidence shows that glacier shrinkage will continue in a warming climate, so contributing to additional sea level rise. The Greenland ice sheet is thought to be reducing overall but there are significant uncertainties due to limited data. While there is evidence of glacier retreat in the southwestern part of Greenland that would contribute to sea level rise, there is also evidence that increased precipitation is contributing to thickening ice accumulation in the interior part of the ice sheet that would act to offset sea level rise.

The situation with the Antarctic ice sheet is also complex, with significant uncertainties. As a result, it is unknown whether the Antarctic ice sheet has contributed to sea level rise over the last 100 years or not. The evidence points to a likely future increase in Antarctic ice accumulation as a result of global warming and increased precipitation. Warrick et al. [Ref 47] summarises the estimated contributions to sea level rise at section 9.4.8 quote;

"Assuming the contribution from Antarctica has been zero, the combined contributions from thermal expansion, mountain glaciers and the Greenland ice sheet over the last 100 years total 105cm. This is within the range of observed sea level rise (10 – 20cm), albeit at the lower end. The range of uncertainty is large -0 5cm to 22cm."

Local sea level

Although global sea level rise is of concern, what is relevant to any specific location and any flood mitigation planning is the local sea level. Distinction needs to be made of the difference between the global average sea level and the local relative sea level (LRSL). The LRSL is determined by the rate of ground movement up or down, the rate of sediment deposit from rivers and the rate of global sea level rise.

In some locations previously under the weight of great northern hemisphere ice caps 20,000 years ago, the ground is still rising after the ice melted. Some locations are rising at a rate of 9mm per year such that the LRSL is falling. In other locations such as major river deltas in the Gulf of Mexico and Bangladesh, the ground is sinking such that a greater rise in LRSL than the average global sea level is being experienced. Some of these locations would be at risk from coastal flooding even if the global sea level was static.

The IPCC provides sea level projections but due to the uncertainties in the science these projections should be treated with caution. The sea level rise projected by the IPCC for the UK coastal areas by year 2100 based on a 3.0°C global temperature rise is around 40 to 70cm depending upon location. Refer to the NASA IPCC AR6 Sea Level Projection Tool [Ref 48] that provides for multiple scenarios and for various global locations.

> The main causes of sea level rise are thermal expansion of the ocean and melting of land-based ice sheets and mountain glaciers due to global warming. However, there is no evidence that the rate of rise is increasing significantly or that human activities are having any measurable effect. Also, given that the time constant for the ocean and its thermal expansion is in the order of 1,000 years, sudden changes to the current trend for a slow rise in sea level is most unlikely. However, the rise in sea level does present a serious risk of coastal flooding in some locations, potentially affecting millions of people worldwide.

Ocean acidification

A discussion on ocean acidification needs to start with understanding what is meant by acid and alkaline (also known as 'basic'). An ion is an atom or molecule that has gained or lost an electron giving it an electrical charge, making it chemically reactive. An acid has gained a number of positive hydrogen ions (H+) and an alkaline has gained a number of negative hydroxide ions (OH-). The pH scale (for *'pondus Hydrogenium'* or power of hydrogen) is used to measure the wide range of acid – alkaline conditions.

The pH scale ranges from 0 (very acidic) through 7 (neutral as in pure water) to 14 (very alkaline). The scale is logarithmic in that the number of hydrogen ions is multiplied by 10 for each pH scale increment. A solution with a pH of 0 has 10 times the ions of a solution with a pH of 1 and 100 times the ions of a solution with a pH of 2 and so on. Refer to the NOAA, A Primer on pH [Ref 49] web page for additional information. The healthy human body is slightly alkaline within the range of 7.35 to 7.45.

The ocean is alkaline and has been so for at least 750 million years, when the atmospheric CO_2 concentration was around 20 times greater than today. Since that time CO_2 has been gradually removed from the atmosphere and

A Guide to the Climate Debate

deposited via carbonate sediment deposits into limestone and other minerals. As a result of this natural buffering ability, it is extremely difficult to permanently change the ocean pH by adding acids. When seawater absorbs CO_2, carbonic acid is produced that will act to reduce the pH value. Additional information can be found on the Yale, Ocean Acidification [Ref 50] web page.

Ocean chemical processes that involve CO_2 are controlled by the saturation level, this being the maximum amount of the gas that can be dissolved in a given volume. The saturation level varies with temperature, pressure and the concentration of other dissolved materials. The CO_2 level decreases with increasing temperature and increases with increasing pressure. Saturation at the ocean surface is controlled mainly by temperature, and at ocean depths below around 3.5km it is controlled by pressure. This results in seawater pH being typically 7.5 at depth and 8.5 in surface waters although it is locally variable.

Although there is a limit on dissolved CO_2, through the process of crystallisation and precipitation, especially in warm shallow tropical ocean water, dissolved CO_2 eventually leads to the accumulation of solid carbonates on the sea floor. Therefore, there is no limit on how much CO_2 can be added to the ocean, the only limitation being the availability of CO_2 in the atmosphere.

The alarmist ocean acidification claims by Warmists would appear to be exaggerated and are based on projections that by definition are unsound and subject to the large areas of uncertainty in the science.

The evidence reviewed shows that a continued high atmospheric CO_2 concentration will have a slight acidification (less alkaline) effect of the ocean's surface waters. With surface seawater pH projected to change to around 7.8 by year 2100, it will still be more alkaline than seawater from the deeper ocean.

A Guide to the Climate Debate

Although acidification can affect marine organisms, there appears to be no clear evidence that this slight change in the pH of surface seawater will have any significant effect globally on marine life. However, there is a recognition that marine life could be affected locally, especially in tropical areas due to surface sea water warming.

Q9. What is the record of climate change predictions?

Climate change predictions have attracted a reputation for being both alarmist and inaccurate. These failed predictions reduce confidence in the abilities of the claimed climate experts and call into question the entire climate change hypothesis. This history of failed predictions and alarmist warnings prepares the ground for a repeat of Aesop's fable about the shepherd boy that cried wolf.

Predictions

During the 1970s there were predictions of global cooling and that the planet was entering a new ice age. Fortunately, this did not materialise and during the 1980s the global cooling predictions were replaced by warnings of global warming. Some of the most notable failed predictions include:

- In 1978 it was predicted that the concentration of atmospheric CO_2 would double by 2020. Levels only increased by 23%.
- In 1987 NASA's James Hansen predicted that the world would be 3°C warmer by 2020. Global Average Temperature was about 0.44°C higher in 2020.
- In 2000 it was predicted that by 2020 snow will be so unfamiliar to people in the UK that they won't know how to deal with it. The reality is it still snows in the UK.
- In 2006, Al Gore the former US vice president claimed that unless world leaders took *'drastic measures'* to reduce greenhouse-gas emissions, the earth would surpass the 'point of no return' in ten years — a 'true planetary emergency,' he called it. Fortunately, it did not happen.

- In 2007, the IPCC predicted the Himalayan glaciers would disappear by 2035. The U.N.'s chief climate science body retracted the claim in 2010, explaining the prediction wasn't based on any peer-reviewed data, but on a media interview with a scientist conducted in 1999.
- In 2008 it was predicted that snow on Mt. Kilimanjaro would vanish by 2020. The snow is still there.
- In 2009 it was predicted that China would cut emissions 40-45% below the 2005 level by 2020; India to cut 20-25%. The reality was by 2020: China's 2020 emissions ~85% higher than 2005; India's ~150% higher.
- In 2009 it was predicted that the glaciers would be gone from Montana Glacier National Park in the US by 2020. The glaciers are still there but what has gone are the signs claiming Glacier National Park glaciers would be gone by 2020.
- Also in 2009, former US vice president Al Gore declared that *"there is a 75% chance that the entire north polar ice cap, during some of the summer months, could be completely ice-free within the next five to seven years."* Then in 2013 an article in the Guardian newspaper claimed an ice free arctic in two years. The ice is still there.

The fact that so many of these warnings have proved to be wrong has not stemmed the flow, as we experience alarmist climate change reports every day in the mainstream media. Some may describe this as propaganda, but there is no doubt that this continuous flow of alarmist messages for over 40 years has increased public awareness and created a widespread belief in man-made climate change. This despite the fact most alarmist reports when examined are found to be factually inaccurate and misleading.

General Circulation Models

The source of many of these alarmist climate change claims are the IPCC climate models. The IPCC climate models also known as General Circulation

A Guide to the Climate Debate

Models (GCMs) use mathematical equations to characterise how energy and matter interact in different parts of the ocean, atmosphere and land. See the IPCC - What is a GCM? [Ref 51] web page.

When fed with appropriate initial data these models calculate possible future climate states. These models are deterministic in that every factor that is known to influence climate must be included from the start. These models therefore assume that we have a complete understanding of the climate system, which of course we don't. As our examination of the evidence shows, the level of uncertainty with each of the climate processes is very high.

Since some climate processes are too complex to be fully represented in the model they are replaced by 'parameterised' relationships that in reality are no more than educated guesses. These parameters can be adjusted and invariably the model output is adjusted to conform with the creator's expectation.

The IPCC will claim that the outputs of their climate models are simply 'projections' for differing scenarios and are not claimed to be climate predictions or forecasts. The GCMs are to all intents and purposes scientific computer games where the value of the GCM is as a valuable learning tool. Given the known shortcomings with such climate models, to use them for forecasting purposes would be either foolish or dishonest. In fact, the IPCC scientists fully recognise that the GCMs are unable to provide climate forecasts which is why they refer to the GCM outputs as 'projections.

An alternative to the deterministic GCMs used by the IPCC is statistical climate modelling. This is a well understood scientific alternative and uses data from past climatic conditions to construct projections of future climate. Although the IPCC appear to ignore this methodology, these models have identified a reoccurring 60-year climate cycle that corresponds with the Pacific Decadal Oscillation (PDO) that resembles a large scale El Nino/La Nino climate oscillation. Several of these models have independently arrived at a similar conclusion by forecasting cooling for the first decades of the 21st century followed by a warming.

A Guide to the Climate Debate

The 2010 paper by Syun-Ichi Akasofu titled; On the recovery from the Little Ice Age [Ref 52] provides further details. So, depending upon the climate model you choose and the input parameters you select, you can generate outputs that demonstrate either global warming or global cooling.

A further discussion on the subject includes the paper Validity of Climate Change Forecasting for Public Policy Decision Making [Ref 53] by Green *et al*. from 2009 where it questions the IPCC forecasting techniques. The IPCC projections were compared with the simpler alternative statistical 'benchmarking' method and found that the IPCC errors with their GCMs were seven times greater than the benchmark method. One has to question why the IPCC continue to use this misleading and inaccurate forecasting methodology?

Meanwhile the IPCC deterministic GCM projections are seized upon by the media and other commentators who mislead themselves and the public with these 'projections' that are invariably interpreted as climate 'predictions' or 'forecasts'. The IPCC fail to correct these misinterpretations so should be considered as complicit with the deception.

> The IPCC has employed some of the best public relations and advertising agencies to help convey the most effective alarmist climate messages possible. We now experience alarmist climate change propaganda on a daily basis from multiple sources and on multiple topics. It seems that whatever is wrong in the world is the result of climate change.

Part of the Warmist / IPCC ploy and a perversion of science is that those who challenge the climate change hypothesis of global warming caused by CO_2 emissions from human activities are expected to be able to provide the proof. Yet to date the Warmists have failed to provide scientific proof of their climate change hypothesis. This despite the spending of many billions of dollars over the last four decades to find scientific proof of man-made global warming.

The 'Hockey Stick' deception

In 1998 'Nature' magazine, published a paper titled, Global-scale temperature patterns and climate forcing over the past six centuries [Ref 54] by Mann, Bradley and Hughes. This was followed in 1999 by a second paper published by Geophysical Research Letters (GRL) titled, Northern hemisphere temperature during the past millennium: inferences, uncertainties, and limitations [Ref 55] again by Mann, Bradley & Hughes. A third paper was published by GRL during August 2003 titled, Global surface temperatures over the past two millennia [Ref 56], this time by Michael E Mann & Philip D Jones.

The Mann et al. papers were based on statistical analysis of tree ring records from locations across the northern hemisphere. Note; tree ring data from ancient trees is used as it provides historical evidence of growing conditions that can be interpreted to provide a proxy of temperature although there are limitations with this methodology. The resulting reconstructed temperatures for the 1st paper covered the period 1400 up to 1900, then projected to 2000 with thermometer temperature records. For the second paper this was extended to cover the last 1,000 years and the third paper extended further to cover the past 2,000 years.

The temperature graph exhibits a slowly declining temperature from around 1200 to 1900, then followed by a rapid rise, refer to Figure 3a [Ref 55]. The curve became known as the infamous 'Hockey Stick' due to its resemblance to a North American ice hockey stick. The Hockey Stick curve from the second Mann et al. paper featured prominently in the IPCC's Third Assessment Report (TAR) during 2001 including the TAR Summary for Policymakers, Figure SPM-10b [Ref 57], reproduced here as Figure 20.

The Hockey Stick curve was of critical importance to the IPCC at that time as it allowed them to claim that dangerous human caused global warming was taking place. By removing the Medieval Warm Period from the record, the IPCC could claim that the modern warming was unprecedented over the last 1,000 years. At this time, the Hockey Stick graph was the most important graph in the world and was subsequently heavily marketed by environmental

A Guide to the Climate Debate

groups. We discussed the Medieval Warm Period earlier at key question 1, a 300-year period almost 1000 years ago when the earth was warmer than today.

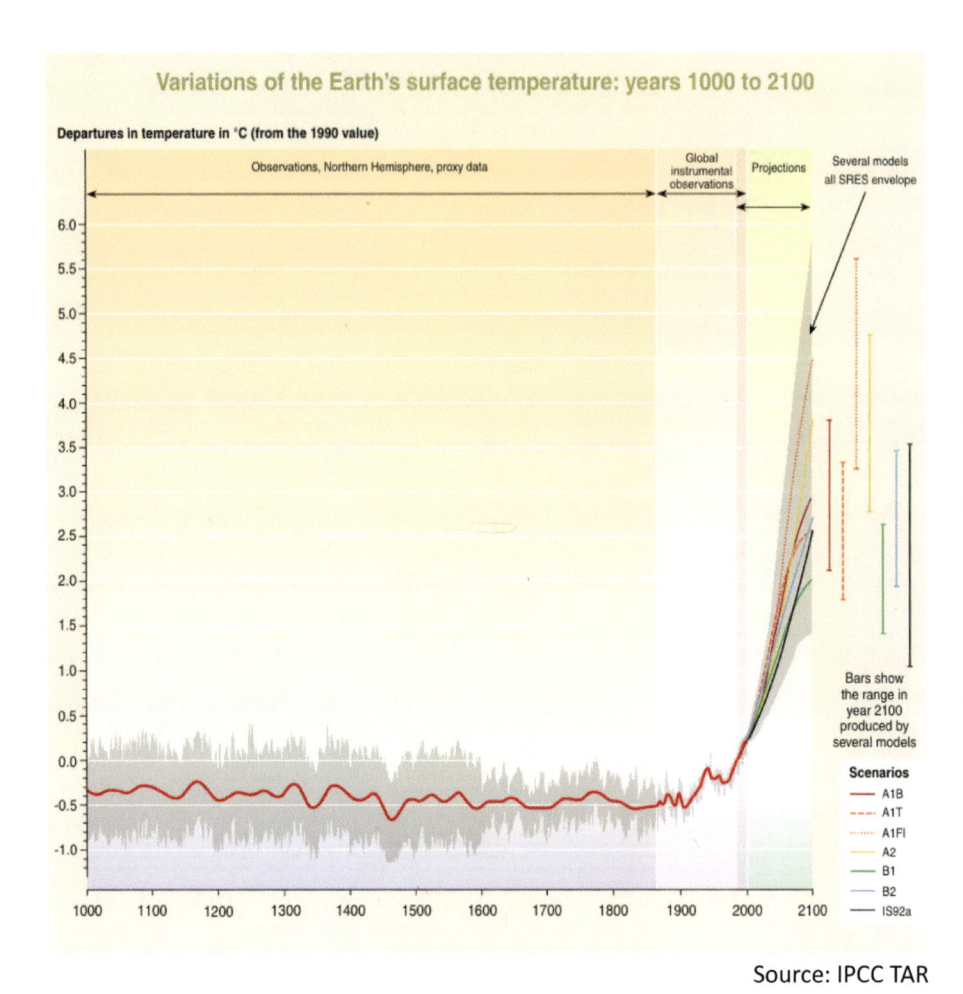

Source: IPCC TAR

Figure 20: Infamous 'Hockey Stick' curve reproduced by the IPCC in the TAR (2001) as Figure SPM-10b showing variations of the earth's surface temperature: years 1000 to 2100

A Guide to the Climate Debate

The Mann *et al.* papers and the Hockey Stick curve were however contradicted by earlier findings in that known climatic periods such as the Mediaeval Warm Period and the Little Ice Age were not shown. Also, the rapid temperature rise shown during the 20th century for the northern hemisphere was not evident in records for the southern hemisphere. Additionally, the IPCC Figure SPM-10b (Figure 20) was made to appear much more dramatic by the inclusion of GCM created temperature projections to 2100. The Mann paper soon became the target of Sceptic scientists who were highly suspicious of the findings.

One of the first to rebut the Hockey Stick paper were Soon and Baliunas in their paper published during 2003 by Climate Research titled Proxy climate and environmental changes of the past 1000 years [Ref 58]. The paper concluded that, *"Across the world, many records reveal that the 20th century is probably not the warmest nor a uniquely extreme climatic period of the last millennium"*, so directly challenging the Hockey Stick papers and the IPCC.

This was followed by two papers by McIntyre and McKitrick who focused on the statistical analysis method used by Mann *et al*. The first titled, Corrections to the Mann et al. (1998) proxy data base and northern hemisphere temperature series [Ref 59] dated 2003 was published by Energy & Environment. This paper also directly challenged the Hockey Stick papers and the IPCC, with the Abstract stating;

"The data set of proxies of past climate used in Mann, Bradley and Hughes (1998, "MBH98" hereafter) for the estimation of temperatures from 1400 to 1980 contains collation errors, unjustifiable truncation or extrapolation of source data, obsolete data, geographical location errors, incorrect calculation of principal components and other quality control defects. We detail these errors and defects. We then apply MBH98 methodology to the construction of a Northern Hemisphere average temperature index for the 1400-1980 period, using corrected and updated source data. The major finding is that the values in the early 15th century exceed any values in the 20th century. The particular "hockey stick" shape derived in the MBH98 proxy construction – a temperature index that decreases slightly between the early 15th century and early 20th century and then increases dramatically up to 1980 — is primarily

an artefact of poor data handling, obsolete data and incorrect calculation of principal components".

The second paper by McIntyre and McKitrick titled Hockey sticks, principal components, and spurious significance [Ref 60] dated February 2005 was published by GRL. This paper further challenged the statistical methodology used by Mann *et al.* to derive the Hockey Stick curve in the strongest terms with the paper Abstract stating;

"The "hockey stick" shaped temperature reconstruction of Mann et al. (1998, 1999) has been widely applied. However it has not been previously noted in print that, prior to their principal components (PCs) analysis on tree ring networks, they carried out an unusual data transformation which strongly affects the resulting PCs. Their method, when tested on persistent red noise, nearly always produces a hockey stick shaped first principal component (PC1) and overstates the first eigenvalue. In the controversial 15th century period, the MBH98 method effectively selects only one species (bristlecone pine) into the critical North American PC1, making it implausible to describe it as the "dominant pattern of variance". Through Monte Carlo analysis, we show that MBH98 benchmarks for significance of the Reduction of Error (RE) statistic are substantially under-stated and, using a range of cross-validation statistics, we show that the MBH98 15th century reconstruction lacks statistical significance".

Several other papers followed, so effectively exposing the sophisticated Hockey Stick deception by Mann et al. along with a complicit IPCC.

The evidence shows that the Hockey Stick graph deceived governments and political leaders across the world when the IPCC TAR was released in 2001 by claiming that global warming was much more significant than the reality.

The book by Andrew Montford published in 2010 titled '**The Hockey Stick Illusion**' [Ref E] provides the definitive story of the Hockey Stick deception and how it all unravelled. Montford's Hockey Stick story starts in 1977 with the

A Guide to the Climate Debate

first World Climate Conference organised by the World Meteorological Organisation (WMO) and the first major recognition of climate change, due to both natural and anthropogenic causes. The conference resulted in a 'Call to Nations', for steps to be taken to improve knowledge of climate and for potential man-made changes to climate to be foreseen and prevented. So, not just a call for more research but for a particular policy outcome. Already the idea was fermenting with some scientists of a potential source of research funding and influence without end.

Over the next few years, the idea of global warming due to the greenhouse effect developed slowly, with the IPCC reporting in its First Assessment Report (FAR) in 1990;

"A global warming of larger size has almost certainly occurred at least once since the end of the last glaciation without any appreciable increase in greenhouse gases. Because we do not understand the reasons for these past warming events, it is not yet possible to attribute a specific proportion of the recent, smaller warming to an increase of greenhouse gases."

The historical evidence presented the global warming movement with a major problem as there was no evidence that the recent small rise in global temperature supported the argument for catastrophic global warming. There is evidence that during the Medieval Warm Period from the eleventh to fifteenth centuries it had been warmer, then followed by the Little Ice Age up to the nineteenth century, much cooler. Since then, the climate has warmed again.

It is common knowledge that during the Medieval Warm Period the Vikings had colonised Iceland and Greenland where they managed to grow crops, and grapes had been grown commercially in England. Then during the Little Ice Age, the Vikings abandoned Greenland and ice fairs were held on the River Thames at London when the river froze over. Refer to the book by Brian Fagan published in 2000 titled; **The Little Ice Age**' [Ref F] for a historical account of this period and how the changing climate influenced history.

It became clear at this time for the warming movement that in order to demonstrate that the current warming trend was unprecedented and hence attract research funding, it was necessary to somehow lose the Medieval Warm period.   Montford documents several attempts to remove the Medieval Warm Period temperature rise from history including the Deming affair on page 27.

Deming is quoted; *"With the publication of the article in Science, I gained significant credibility in the community of scientists working on climate change. They thought I was one of them, someone who would pervert science in the service of social and political causes. So one of them let his guard down, A major person working in the area of climate change and global warming sent me an astonishing email that said, 'We have to get rid of the Medieval Warm Period'."* The sender of the email was later identified as Jonathan Overpeck of the University of Arizona. Overpeck was subsequently one of the lead authors for the IPCC Fourth Assessment Report (FAR) in 2007. It became apparent that a concerted effort was made to rewrite the earth's climate history such that the Medieval Warm period disappeared.

Progress with eliminating the Medieval Warm Period and changing the narrative become more apparent with the IPCC Second Assessment Report (SAR) in 1995 that stated;

"Based on the incomplete observations and paleoclimatic evidence available, it seems unlikely that global temperatures have increased by 1°C or more in a century at any time during the last 10,000 years".

And;

"The limited available evidence from proxy climate indicators suggests that the 20th century global mean temperature is at least as warm as any other century since 1400 AD. Data prior to 1400 are too sparse to allow the reliable estimations of global mean temperature."
This effectively claiming that the modern warming is unprecedented.

A Guide to the Climate Debate

All this paved the way for the Hockey Stick paper published by Michael Mann in 1998 [Ref 54], with the second paper in 1999 [Ref 55]. In these papers Mann produced the infamous (ice) hockey stick shaped curve that completely loses the Medieval Warm Period and the Little Ice Age. The curve was created from a combination of tree ring data analysis, the hockey stick handle, plus more recent instrument measurements, the hockey stick blade. This curve was subsequently reproduced in the IPCC Third Assessment Report (TAR) in 2001 with the addition of GCM temperature projections extending the curve to year 2100, see Figure 20.

This dramatic graph enabled the IPCC to claim that the modern warming was unprecedented in the last 1,000 years and was the result of human activity. In 2001 this graph aimed at governments was the most influential graph in the world significantly influencing public opinion into believing in catastrophic climate change. The Medieval Warm Period having been effectively erased from history. Note that Mann was also a lead author for the IPCC TAR where he subsequently assessed his own work, so a clear conflict of interest.

Montford proceeds to describe the Hockey Stick paper in forensic detail and the associated unravelling of the complex statistical analysis by the Canadian mathematician Steve McIntyre and a colleague Ross McKitrick. The Mann paper is based on the analysis of tree rings based on the assumption that the tree rings can act as a proxy for the global temperature. The tree ring data is obtained from cores drilled from the trunks of ancient trees. However, tree growth and hence tree ring data does not always follow temperature, especially in arid locations where rainfall is the limiting factor. He also discusses the practice of 'cherry-picking' where scientists select certain trees or groups of trees from certain locations in order to obtain the result that best suits their objectives.

McIntyre discovered that the selective use of just a very small number of these proxies (tree ring samples that are claimed to capture temperature information) were able to produce a hockey stick shaped temperature reconstruction. In fact, the hockey stick shape is only provided by a very small number of Bristlecone pine trees located in the western USA. Much of Montford's book discusses the statistical analysis methods used by Mann and his supporters and how novel analytical methodologies were applied such that irrespective of the raw data, whether real tree ring data or simply random noise data, the temperature reconstruction would always be a hockey stick shape.

Later tree core sampling work revealed that the growth spurt exhibited during the late 20th century by these bristlecone pine trees was due to the uneven growth due to the 'strip bark' effect, and not as a result of rising temperature. With the tree ring data not providing a proxy for global temperature, the hockey stick temperature reconstruction curve was shown to be simply a product of the statistical methodology employed in their analysis.

Montford also shines a light on the IPCC, its role in the Hockey Stick saga and an especially critical appraisal of its review process for the Assessment Reports. The IPCC protocols allowed Mann to be appointed as the lead author on the paleoclimate chapter where he was responsible for reviewing his own work, forming an opinion on the rest of the scientific inputs and writing the final text. Montford states; *"this situation would have been entirely unacceptable in a commercial situation, and in fact would have been entirely illegal outside a banana republic"*.

Montford goes on to describe in detail in chapter 11 the corrupt process within the IPCC where scientific papers were selected and reviewed for the Fourth Assessment Report (FAR) allowing continued prominence to the Hockey Stick despite the mounting evidence that the Hockey Stick curve was not a credible reconstruction.

At the heading to chapter 15, Montford quotes Atte Korhola, Professor of Environmental Change, University of Helsinki;

> *"When later generations learn about climate change science, they will classify the beginning of the twenty-first century as an embarrassing chapter in the history of science. They will wonder about our time and use it as a warning of how the core values and criteria of science were allowed little by little to be forgotten, as the actual research topic of climate change turned into a political and social playground."*

Montford then questions how the Hockey Stick succeeded as it did despite having got the millennial temperature reconstructions so wrong? It was a peer-reviewed paper published in one of the world's most prestigious scientific journals. It also passed another detailed review before appearing in the IPCC's Third Assessment Report.

If we examine the peer review process, prior to the mid-twentieth century, peer-review was rarely used, but since then nearly all scientific papers are peer-reviewed. When analysed, Montford finds that peer-review does not actually achieve very much. It seems to benefit mainly the journals themselves and certainly does not mean a peer-reviewed paper is correct. He quotes Richard Smith, a former editor of the British Medical Journal;

"We have little evidence on the effectiveness of peer review, but we have considerable evidence of its defects. In addition to being poor at detecting gross defects and almost useless for detecting fraud, it is slow, expensive, profligate of academic time, highly subjective, something of a lottery, prone to bias and easily abused."

The most secure method of verification of scientific findings is by replication. Only through reproducing the claimed findings of a scientific paper can other researchers be certain that those claimed findings are correct. Over the last one hundred years as science has advanced and where computers with vast datasets dominate scientific research, the paper today is not the scholarship, it is merely the advertising of the scholarship. The actual scholarship is the data and code used to generate the paper's findings; hence replication of the findings is only possible if this data and code are made available.

As was shown with the Hockey Stick and other related papers, attempts to obtain the data and code were met with a wall of evasion and obfuscation by the paleoclimate community, so preventing any independent reproductions and hence verification of the findings.

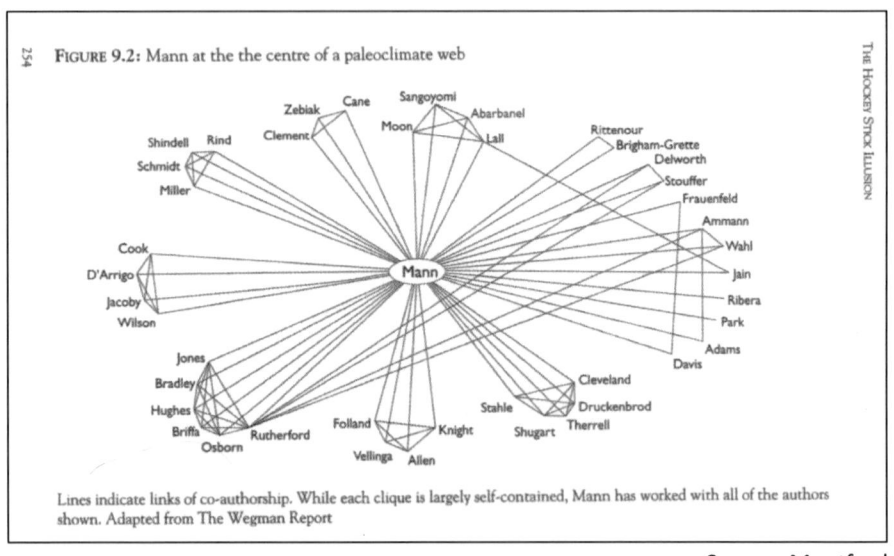

Figure 21: Mann at the centre of a paleoclimate web

Source: Montford

Earlier in his book at chapter 9, Montford discussed the paleoclimate community involved with Mann and the Hockey Stick. An analysis by the researcher Wegman (page 252) into this group of paleoclimate researchers found them to be too insular, too self-contained, and too close knit, with Mann positioned at the centre. This effectively prevented any independent review of Mann's work. This chart shown by Montford as Figure 9.2 is reproduced here as Figure 21.

While refusing to release the data and code would be regarded by most reasonable people to be unethical, in the case of publicly funded research that has enormous government policy implications, such behaviour should be totally unacceptable.

Climategate

During November 2009 over 1,000 emails were released into the public domain from the servers of the Climate Research Unit (CRU) by either a hacker or a whistleblower. The CRU is part of the University of East Anglia (UEA), and an important contributor to IPCC climate reports. The leak including 10 years' worth of emails, data and code soon became known as 'Climategate'. Whereas computer hacking is an illegal activity, release of the data by a whistleblower could be considered legal under UK law, although this was never tested as the source of the email exposure was never established.

The leaked emails exposed the widespread practice by some of the world's leading climate scientists of data manipulation in order to obtain the best results that support the pro-anthropogenic global warming agenda. It also provided evidence of hiding and destroying data to obstruct Freedom of Information (FOI) laws, manipulation of the peer-review process, threats to scientific journals (to toe the line) and attempting to silence dissenting views.

The emails also revealed a desire by these climate scientists to get rid of the Medieval Warm Period from the data records as this was proving to be an inconvenience to the Warmist climate change narrative. Included in the leaked emails was correspondence from those scientists responsible for the hockey stick deception as already discussed. These revelations finally confirmed the claims of Sceptics regarding the Hockey Stick deception that took place a decade earlier.

A second book by Andrew Montford published in 2012 titled '**Hiding the Decline**' [Ref D] provides the definitive story in forensic detail of 'Climategate' including the subsequent investigations. '**Hiding the Decline**' is the sequel to his earlier book '**The Hockey Stick Illusion**' [Ref E].

The leaked emails also described how the temperature graph SPM.1 (reproduced here as Figure 22) from the IPCC Fourth Assessment Report (AR4) [Ref 61], published during 2007 was also manipulated to show a 'clean' temperature rise to support the IPCC climate change hypothesis. It was this manipulation of the data and the hiding of data showing a fall in temperature, the so-called 'trick' that gave the name '**Hiding the Decline**' to Montford's book.

One of the most important revelations from Climategate was that the UEA scientists had lost or destroyed the raw temperature data on which their claims of global warming are based. This was data that Sceptics had been requesting unsuccessfully for years in order to verify the Warmist claims.

This data collected from weather stations around the world for the last 150 years was then 'adjusted' to take account of 'variables' including the UHI effect, in the way the data was collected. The corrected (adjusted) data had been retained but the original data destroyed, so it's now impossible to check these global warming claims. The Sceptic's concerns with temperature data and corrections to account for the UHI effect have been discussed earlier at key question 1.

Climategate led to the resignation of the CRU Director, Professor Phil Jones who was shown to be at the heart of Climategate, and led to several formal investigations.

A Guide to the Climate Debate

Changes in temperature, sea level and Northern Hemisphere snow cover

(a) Global average surface temperature

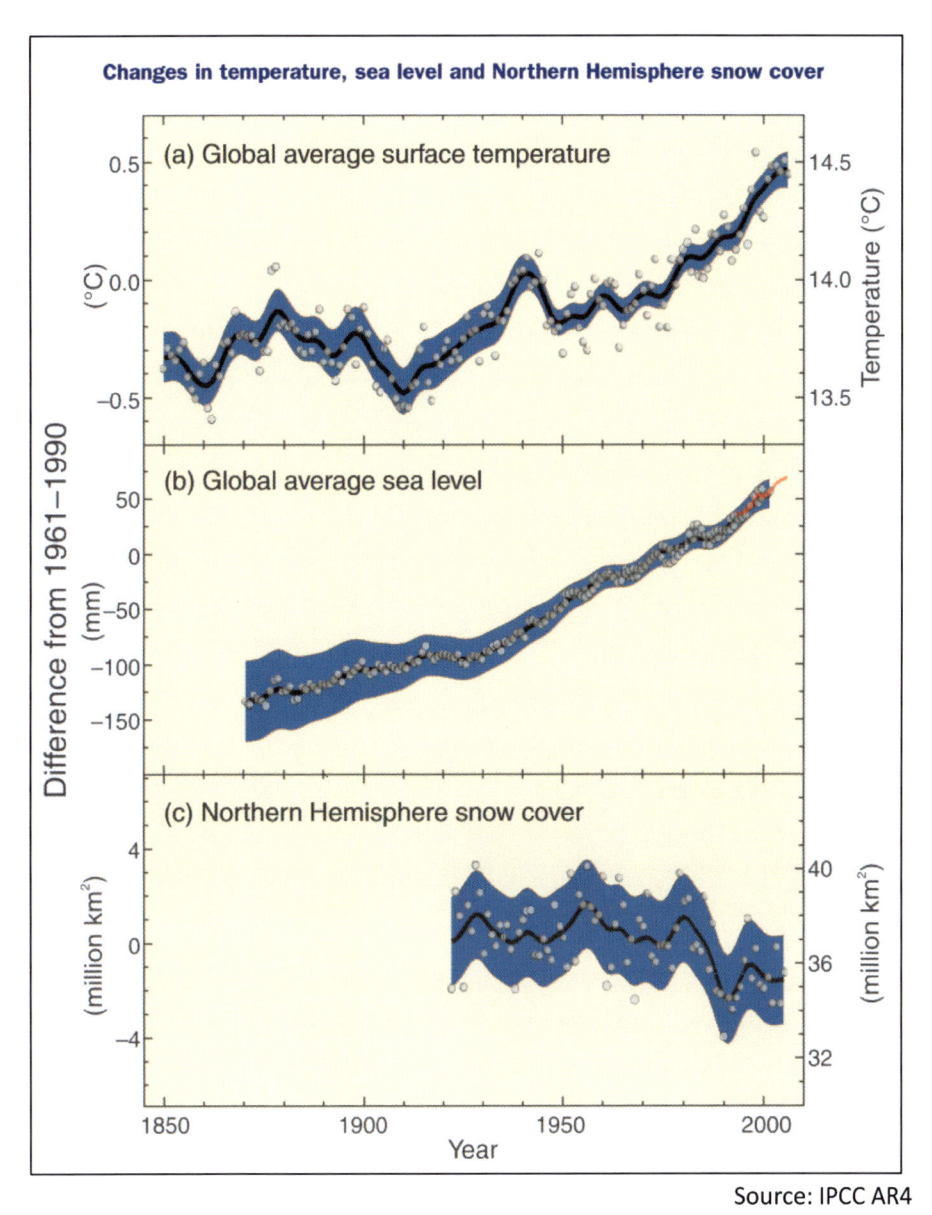

(b) Global average sea level

(c) Northern Hemisphere snow cover

Source: IPCC AR4

Figure 22: AR4 Figure SPM.1 shows the smoothed surface temperature curve manipulated using the 'trick' to show a 'clean' temperature rise.

These Climategate investigations included:

- The UK House of Commons Science and Technology Select Committee chaired by Phil Willis. Their report was released during March 2010.

- Penn State University conducted their own investigation in the US into the activities of their climate scientist Michael Mann. Mann was a central figure in the Hockey Stick deception with his activities exposed by the Climategate email release. This two-stage inquiry was carried out behind closed doors and the investigatory committee released its report during June 2010.

- The UEA commissioned their own investigation chaired by Sir Muir Russell. This investigation subsequently changed its brief to focus only on any misconduct and to exclude the science. An investigation into the science was subsequently delegated to a separate investigation. The UEA Russell investigation released its report during July 2010.

- This separate investigation by a Science Assessment Panel into the science behind Climategate was instigated by the UEA. It was conducted by an independent panel chaired by Lord Oxburgh but this investigation also managed to avoid investigating the science. The Oxburgh report was published April 2010.

These investigations resulted in no individuals being found guilty of anything other than minor infringements, no one was to blame for the deceptions and the chief protagonists Jones and Mann were cleared. This is a most remarkable conclusion given the weight of evidence. Phil Jones was subsequently reappointed by the CRU and even today the Climate Research Unit [Ref 62] continues to deny any wrongdoing and hails Phil Jones as a hero.

A Guide to the Climate Debate

However, the information from the emails catalogued by Montford in 'Hiding the Decline' showed clear evidence of manipulation of data, manipulation of the peer-review process, and attempts to silence climate change dissent. There was clear evidence of obstruction of FOI laws including the hiding and deletion of data, but due to the 6-month Statute of Limitations applying, no individuals were ever charged.

These investigations as described by Montford, were taken over by Establishment figures who were selective in the evidence they reviewed, the individuals they interviewed and the questions they asked. The outcomes were widely anticipated by Sceptics and with all of these investigations, there was an impression that the intentions all along had been to protect the integrity of the climate science community, especially the reputation of the CRU and the IPCC, and critically to steer clear of the science, so avoiding exposing any cracks in the climate change hypothesis. Many considered these investigations as an Establishment whitewash.

The conclusions that Montford comes to after discussing the Hockey Stick deception Ref[E] and the subsequent Climategate emails Ref[D] are:

- *"The scientific literature is no longer a representation of the state of human knowledge about climate. It is a representation of what a small cabal of scientists feel is worthy of discussion.*
- *The IPCC reports represent the outcome of a process in which a relatively small group of scientists produce a biased review of a literature they themselves have colluded to distort through gatekeeping and intimidation.*
- *The released emails establish a pattern of behaviour that is completely at odds with what the public has been told regarding the integrity of climate science and the rigour of the IPCC report writing process. It is clear the public can no longer trust what they have been told."*

What did emerge from these investigations is that there is no system of accountability to ensure scientists' good behaviour and the peer review process is shown to be inadequate. With tens of thousands of scientists in the UK, there have not been any convictions for research fraud in recent years.

A Guide to the Climate Debate

Q10. Why has climate change become political?

> During the early seventeenth century the scientific consensus among astronomers was that the sun circled the earth. At that time a 'Sceptic' by the name of Galileo who presented scientific evidence to the contrary, was ridiculed, imprisoned and tried for heresy.

How could a similar scientific dispute, be repeated with similar results in the 21st century'? To answer this, we need to examine how big politics and big business has taken over the climate change narrative, and how it has been achieved over four decades. During this time the climate change narrative has evolved into what has become a new religion known as 'climatism'.

A Noble Cause

So how could such a corruption of science and indoctrination of the general public take place and on such a scale? Surely scientists are ethically bound to scientific principles, and most definitely are, but all are human beings as well. Like most of us they will also want to see the world become a better place while at the same time pursue a successful career and care for their families. As a result of a perceived moral righteousness, some scientists have strayed from their strict scientific discipline and have become advocates of the global warming cause. This corruption of the science was demonstrated by the revelations from the Climategate release of emails.

On examination it is not too difficult to see how climatism has evolved and taken hold globally. What started out as a theory during the 1970s and 80s by a small number of climate scientists that just maybe, a rising atmospheric concentration of the minor greenhouse gas CO_2, could cause warming of the atmosphere was taken up by environmentalists, eventually becoming a belief. The attraction of man-made CO_2 emissions as the villain to the new breed of mostly left-wing environmental extremists is that CO_2 emissions are a byproduct of burning fossil fuels, so a symbol of global capitalism. By this time a scientific theory was already becoming a political cause.

The article appearing on the Net Zero Watch website by Richard Lindzen dated 20 March 2024 titled; How did the obsession with decarbonization arise? [Ref 63] provides a discussion on this subject from a USA perspective.

Later as scientific evidence that conflicted with this new ideology was uncovered, some scientists believing the 'noble cause' had become a priority for saving the planet, tended to oppose the new evidence, continuing their support of the anthropogenic climate change, CO_2 theory. Once prominent environmentalists such as Al Gore and other politicians joined the crusade and made government funding available for research into the global warming theory, it became a game changer. The United Nations (UN) soon became a supporter, creating the Intergovernmental Panel on Climate Change (IPCC) during 1988, devoted to promoting the theory of anthropogenic climate change. Within 20 years climatism had taken hold across much of the western world.

It is not just individual scientists that are attempting to control the climate change debate by the use of selective science. Scientists who work for major governmental science agencies are almost all under strict instruction as to the comments that they may make on climate change. In turn governments are also influenced by 'noble cause' corruption where 'saving the planet' is an effective means of attracting green votes. This applies to both Labour and Conservative governments in the UK since the late 1980s.

Governments and politicians have thus arrived at their views and subsequent policies in response to their electorates. These electorates have arrived at their views in response to a long and sustained propaganda campaign by environmental organisations creating a self-reinforcing cycle. As prominent establishment figures such as David Attenborough and King Charles became convinced and supportive of the cause, the self-reinforcement is complete.

A Guide to the Climate Debate

In this current political landscape, no political party could expect to be voted into government today in the UK by advocating an end to the fight against climate change.

If you ask any non-scientist whether they agree with the climate change hypothesis and the need to reduce carbon emissions, almost will respond with, "*Although I do not understand the science, I believe in man-made global warming and support government action to save the planet*", or something similar. Many will respond with incredulity that you could even doubt the so-called experts, so effective has been the propaganda and the widespread conversion to climatism. Images of wild fires, floods, heat waves and collapsing glaciers are seen on our news feeds and television screens on a daily basis continuing to drive home the message that the earth is doomed unless we act now.

If you ask someone who has recently installed solar panels or a heat pump to their house or bought an electric car for their justifications for this extra spending, most will state, "*I wanted to do my bit for the environment*" or similar. Most of these people are well intentioned and no doubt feel some guilt as having personally contributed towards the claimed climate emergency. A high level of individual guilt has been instilled into the population through the decades of alarmist propaganda to motivate individuals to support action now.

Since the formation of the IPCC by the UN in 1988 it has evolved into an inflated political and pseudo-scientific organisation at the heart of the climate change movement and climate change related business interests. There appears no shortage of funding as the IPCC, environmental NGOs and governments employ armies of media, public relations and advertising agencies supported by a compliant media to maintain the flow of alarmist messages.

> Today almost everything that is wrong in the world can be shown as having climate change as an underlying cause, simply because someone has been funded to research that particular topic to look for any connections with climate change. The result is a constant daily flow of alarmist messages justifying the need to reduce carbon emissions. The underlying message is, we must eliminate man-made CO_2 emissions to save the planet and to fix these problems with the world, problems that you have help to create through your use of fossil fuels.

Many of those actively arguing to reduce or eliminate global CO^2 emissions either do not understand the science or don't want to understand it as they consider it irrelevant. They consider the political cause far more important, hence their belief is that *'the ends justify the means'*. In the case of the core of climate scientists, IPCC officials and other establishment figures wedded to the climate change hypothesis, they are in effect playing God.

The debate has long moved from the science to the political arena. Hence the response to suppress any scientific challenge to the climate change hypothesis and the declarations of the scientific 'consensus' and that 'the science is settled'.

Intergovernmental Panel on Climate Change (IPCC)

The IPCC is the United Nations (UN) body for assessing the science related to climate change. It was established in 1988 under the co-sponsorship of the World Meteorological Organisation (WMO) and the United Nations Environmental Program (UNEP), with a charter that directs it to assess peer-reviewed research relevant to the understanding of the risk of anthropogenic climate change (human induced climate change).

The IPCC is an organisation of governments that are members of the UN or WMO and currently has 195 members. See the IPCC History web page [Ref 64] for further information. The IPCC operates in close relationship with the

A Guide to the Climate Debate

United Nations Framework Convention on Climate Change (UNFCCC), see United Nations web page [Ref 65].

The UNFCCC is an international treaty signed in 1992 to combat, quote;

"dangerous human interference with the climate system". As a result, the IPCC defines climate change as *"a change in the climate attributed directly or indirectly to human activity that alters the composition of the global atmosphere and which is in additional to natural climate variability observed over comparable time periods"*.

The IPCC provides regular assessments of the science of climate change, its impacts and future risks, and options for adaptation and mitigation.

The IPCC has to date delivered six Assessment Reports, these are designated:

- First Assessment Report (FAR) during 1990
- Second Assessment Report (SAR) during 1995
- Third Assessment Report (TAR) during 2001
- Fourth Assessment Report (AR4) during 2007
- Fifth Assessment Report (AR5) finalised between 2014 and 2018
- Sixth Assessment Report (AR6) [Ref 66] during 2023

Consulting the most recent report, the AR6, Summary for Policymakers (Governments) opens with the statement on bold text,

"Human activities, principally through emissions of greenhouse gases, have unequivocally caused global warming, with global surface temperature reaching 1.1°C above 1850-1900 in 2011-2020."

The IPCC authors would appear to be confident in this statement even though none of the IPCC reports has been able to claim with scientific proof that global warming due to human activity has been identified and measured, for the very good reason that it has not.

A Guide to the Climate Debate

Many of the IPCC authors are not scientists but include geographers, sociologists and economists, so in effect the IPCC is a marketing organisation summarising the latest evidence, massaging the science, communicating the message and offering advice to governments.

As a result, the global warming discussions in the public domain have ceased to be scientific many years ago and ever since the claims that *'the science is settled'*, and *'there is now a consensus'*. Climate change debate is now mainly about global politics and very big business interests.

The IPCC is heavily influenced by lobbying by two types of pressure groups:

- The environmental groups and NGOs, including World Wide fund for Nature, the Sierra Club, Friends of the Earth and Greenpeace.
- Enterprises whose business depends on the public money thrown into saving the planet. These include renewable energy, biomass, carbon trading and finance.

The IPCC hold numerous conferences and meetings on a regular basis, the most familiar is the annual 'Conference of the Parties' (COP). These are major events that dominate the news each year, with thousands of delegates including government ministers, officials, business leaders and journalists travelling to these conferences. COP28 held at Dubai during 2023 attracted over 65,000 delegates so one can only wonder at the size of the carbon footprint of such an event. The next conference to be known as COP29 is scheduled for November 2024 at Baku, Azerbaijan.

The most recent COPs are:

- COP26 – October 2021, Glasgow, UK
- COP27 - November 2022, Sharm el-Sheik, Egypt
- COP28 - December 2023 Dubai, UAE

The IPCC web site makes clear that it does not conduct scientific research itself but acts to review and summarise scientific papers produced by others.

A Guide to the Climate Debate

A review of the IPCC website illustrates clearly that this is primarily a political organisation with science as a secondary function.

There is no appetite by the IPCC for consideration of alternative scientific evidence on the causes of climate change or even the benefits arising from climate change. Belief in the climate change hypothesis is essential for the continued existence of the IPCC and its dependent organisations. Hence the need to silence any critique of the science and to maintain an atmosphere of mystery, fear and guilt.

Scientific dissent

The intimidation of climate scientists in the UK, the USA and other countries is very real today with platforms for dissenting scientists to publish their findings and the provision of research funding effectively closed off. In the UK the evolution of climatism and adoption of the theory of anthropogenic climate change by the 'establishment' including the main political parties, civil service, the BBC, commercial television, most newspapers, the Royal Society, the Archbishop of Canterbury, David Attenborough and King Charles has left few avenues for the climate sceptic to present conflicting findings or to find a receptive audience. It would be a brave scientist that stands up to these bastions of the establishment and public opinion.

> Of particular concern is the Royal Society that as a scientific academy, has betrayed all that science stands for including the *'scientific method'* by supporting the idea that scientific truths should be established by a political process of consensus, and in suppressing dissent. This takes us back to the experience of Galileo in the 17th century that many assumed could never be repeated today, but as we have seen, has been repeated.

The Royal Society web page Royal Society - How do scientists know [Ref 67] makes it clear that the evidence for climate change being caused by human activities has been verified through the use of climate models. As already discussed, these GCM climate models are based on a limited knowledge base of how the earth's climate system works. Due to the large elements of uncertainty, the use of these GCM climate models is only suitable for projections to aid research. They are simply not able to predict the future climate let alone prove the climate change hypothesis.

The Royal Society News 12 March 2018 [Ref 68] web page claims their Consensus Statement on Climate Change represents the views of tens of thousands of scientists, so confirming their betrayal of what science stands for. As a reminder for the reader, science is based on facts; politics is based on popular opinion.

> The experiences of dissident scientists are discussed in Climate: The Movie [Ref 1] where several scientists discuss their treatment and ostracization for challenging the climate change consensus.

In the face of this growing climate change ideology, individual scientists are faced with a dilemma if they have concerns with aspects of the science. These scientists need to secure research funding in order to continue their career and to support their families. In many countries, scientists that challenge the global warming hypothesis will see their funding disappear, their reputation destroyed and their career ending.

The pragmatic scientist will simply go along with the deception and most have. During this 'investigation of the evidence' some examples of likely 'closet Sceptics' conducting their science in a Warmist science environment have been identified.

Climate change activists realised early on that in order to control the narrative it was necessary to control the funding. The catchphrase that comes to mind is *"Follow the money"* from the 1976 movie titled 'All the president's men', depicting the Watergate political scandal. One of the early proponents and enablers of government funding to research climate change was Al Gore, the former US vice president.

In the UK this control of the narrative is now absolute and during this review of evidence it was not possible to find any recent dissenting scientific opinions from the UK. A similar situation can be found in some other western countries. The dissent that remains is by older, mostly retired scientists who tended to keep their heads down during the early part of their career.

Dissident scientists are often accused of bias due to some receiving research funding from the fossil fuel industry. With the control of government funding firmly in the hands of the Warmist establishment, there is no alternative for the Sceptic scientists if they wish to continue their research. Whereas funding of Sceptic scientists by the fossil fuel industry brings accusations of bias, the vast funding by governments for anthropogenic climate change research doesn't. What matters is who is right, so it should be down to the science, transparency and whether it can stand up to proper scrutiny in line with the *'Scientific Method'*.

Warmist and Sceptic blogs

In the background to the mainstream environmental organisation websites, several referenced in this evidence review, there are blog websites covering both sides of the climate change debate. Some are focusing on the science and rebuttals with some simply focusing on discrediting the individual Sceptics.

Warmist blog sites include:

- Possibly the most infamous blog is [DeSmog](#) [Ref 69], formerly the 'DeSmogBlog' and aimed at what is claimed as climate change disinformation. A review of the web site shows it focuses on the political side of climate change targeting individuals or organisations that could be a threat to the climate change hypothesis.

- A site that appears to be allied with DeSmog is Skeptical Science [Ref 70]. This web site has the stated mission to; *"debunk misinformation that is harming our species' ability to deal with climate change caused by excessive anthropogenic greenhouse gas emissions. We do this by presenting peer-reviewed science and explaining the techniques of climate science denial, discourses of climate delay, and climate solutions denial".*
- The leading Warmist blog website is Real Climate [Ref 71] and this acts an unofficial mouthpiece of the IPCC scientific community. A leading contributor is Michael Mann, the chief architect of the 'Hockey Stick' deception.

Sceptic blog sites include:

- The blog site Climate Audit [Ref 72] edited by Steve McIntyre who was partially responsible for unravelling and exposing the Hockey Stick deception.
- Another Sceptic blog site is Watts Up With That? [Ref 73] edited by Anthony Watts from the USA.
- A UK based website is Net Zero Watch [Ref 74] founded by Nigel Lawson that claims, *"Net Zero Watch highlights the serious implications of expensive and poorly considered climate change policies."* Net Zero Watch published a review of 'Climate the Movie' [Ref 1] by Dr David Whitehouse on their website at Net Zero review of Climate the movie [Ref 75].

Q11. What are the implications of Net Zero policies?

First, we will examine the evolution of Net Zero policies, then look at the impact they will have on reducing CO_2 emissions and on controlling the climate.

The evolution of Net Zero

The drive to reduce greenhouse gas emissions has been driven and coordinated by the United Nations. The first major step was the Kyoto Protocol that was signed in December 1997 by 192 countries and went into force from February 2005. This was an agreement among developed nations to reduce CO_2 and other greenhouse gas emissions in an effort to minimize the impacts of climate change. The emissions target was a modest reduction of 5% below 1990 levels although the EU countries adopted an 8% reduction.

The Doha Amendment signed December 2012 extended the Kyoto Protocol for a second commitment period from 2013 up to 2020. This second commitment period included binding targets for 37 countries including the UK and the European Union to achieve an 18% emission reduction by 2020 compared to 1990 levels. This second commitment was not adopted by the USA, Russia, China, India or the developing countries.

The United Nations then followed with the Paris Agreement signed December 2015 by 196 parties (countries plus the EU), and taking effect from November 2016. The aim of the Paris Agreement is to limit the rise in global surface temperature to less than 2°C above pre-industrial levels and preferably below 1.5°C with the parties committed to reducing greenhouse gases including CO_2.

A Guide to the Climate Debate

During 2017 the Trump administration withdrew the USA, the second largest emitter of greenhouse gases after China, from the Paris Agreement. However, on the first day of the new Biden administration on 20 January 2021, President Biden reversed the Trump decision and the USA was readmitted.

The Paris Agreement is to be implemented by each country via their own national policies in order to reduce their own greenhouse gas emissions. The targets are a 50% reduction of emissions by 2030 and by 2050 the greenhouse gas emissions caused by human activities would need to be reduced to what has become known as Net Zero. Each country that is a party to the agreement was invited to outline their post-2020 emission reduction strategies, known as Nationally Determined Contributions (NDCs).

Countries that signed the Paris Agreement but have not agreed to meet the Paris Agreement climate goals for emission reductions, effectively opting out, include; China, India, Russia, Indonesia, Iran, Libya, Yemen, Mexico, Saudi Arabia, Argentina, Thailand, Singapore, Turkey, Vietnam, New Zealand, South Korea and Canada. Importantly the NDCs are not legally binding on any country, whether or not they currently intend to meet the Paris Agreement goals. Additionally, any party (country) is able to opt out of the Paris Agreement at any time.

In response to the Paris Agreement, in 2019 the UK became the first major economy to adopt a binding (in UK law) Net Zero policy. Then in 2021 the UK government followed with an even more ambitious policy to fully decarbonise UK power generation by 2035. This means eliminating fossil fuels such as coal, oil and gas from power generation. The Labour government elected during July 2024 has now committed to decarbonisation of UK power generation by 2030.

From an electrical engineering perspective, decarbonisation of the UK electricity system by 2030 or even 2035 is considered to be technically and physically impossible to achieve without regular power outages and all that entails for society. The financial and environmental impacts associated with the transformation to renewables is only just starting to take effect, yet we have already seen huge increases in electricity prices.

While the renewables industry repeats the message that renewables are the lowest cost generation, the reality is that renewables are not economically viable without large subsidies and government intervention. Although the war in Ukraine caused a short-term price hike in energy costs, the underlying reason for these price increases for electricity is the cost associated with the transition to renewables and the green levies.

The challenges of replacing fossil fuels with renewable for the generation of electrical power are described by Michael Shellenberger in the TED Talk titled: Why renewables can't save the planet [Ref 76]. While the 17-minute Shellenberger talk is focused on the USA it is equally relevant to the UK.

For the Paris Agreement to succeed in reducing CO_2 emission, the majority of countries and all the major greenhouse gas emitting countries would need to be fully committed, and to actually implement Net Zero policies. However, what we see is that only a minority of countries are making legal commitments, and some of the largest emitters including China and India have only made token gestures of reductions.

It is therefore difficult to imagine how a global Net Zero for emissions can ever be achieved given the disparity of conflicting national interests. For the UK with a 0.9% share of global greenhouse gas emissions, any reductions made will be insignificant and will be eliminated by the rising emissions from China with a 25.8% share, India with a 6.7% share and others. See the Climate Watch website page titled, Explore Long-Term Strategies (LTS) [Ref 77].

The impact of Net Zero

The report titled: Halfway Between Kyoto and 2050 [Ref 78] released 2024 by the Fraser Institute and authored by the Canadian Professor Vaclav Smil, evaluates past carbon emission reduction and the feasibility of eliminating fossil fuels to achieve net-zero carbon by 2050. The report executive summary (with a Canadian bias) states the following:

- *"Despite international agreements, government spending and regulations, and technological advancements, global fossil fuel consumption surged by 55 percent between 1997 and 2023. And the share of fossil fuels in global energy consumption has only decreased from nearly 86 percent in 1997 to approximately 82 percent in 2022.*

- *The first global energy transition, from traditional biomass fuels such as wood and charcoal to fossil fuels, started more than two centuries ago and unfolded gradually. That transition remains incomplete, as billions of people still rely on traditional biomass energies for cooking and heating.*

- *The scale of today's energy transition requires approximately 700 exajoules of new non-carbon energies by 2050, which needs about 38,000 projects the size of BC's Site C or 39,000 equivalents of Muskrat Falls.*

- *Converting energy-intensive processes (e.g., iron smelting, cement, and plastics) to non-fossil alternatives requires solutions not yet available for large-scale use.*

- *The energy transition imposes unprecedented demands for minerals including copper and lithium, which require substantial time to locate and develop mines.*

- *To achieve net-zero carbon, affluent countries will incur costs of at least 20 percent of their annual GDP.*

- *While global cooperation is essential to achieve decarbonization by 2050, major emitters such as the United States, China, and Russia have conflicting interests.*

- *To eliminate carbon emissions by 2050, governments face unprecedented technical, economic and political challenges, making rapid and inexpensive transition impossible."*

A Guide to the Climate Debate

The book titled *'Climate: The Great Delusion'* by the engineer Christian Gerondeau [Ref B] published in 2010, examines the climatic, economic and political realities of climate change. Although now somewhat dated this book still offers some highly relevant observations and common-sense arguments.

Instead of delving into the scientific arguments of climate change, Gerondeau instead starts with the general question: "*what is our room for manoeuvre with regard to CO_2 emissions and is there anything we can do about it?*" Gerondeau opens the discussion by asking five questions and provides simple common-sense answers, reproduced below.

Question 1 - *Is it possible to imagine that mankind would leave any oil, natural gas or coal that could be economically exploited in the planets underground?*
Answer - *The answer is no. The available reserves are not infinite and humanity will look to exploit the world's oil deposits, gas fields and coal mines to the very last drop, cubic metre and tonne economically reachable.*
Question 2 - *Is it possible to stop mankind's use of oil, natural gas and coal resulting in the release of CO_2 when burned?*
Answer - *The answer is also no, for both physical and economic reasons. In most cases there are no technical means of preventing the CO_2 produced being released into the atmosphere. And if a technical solution for sequestrating the CO_2 produced in electric power plants or factories was one day perfected, the very high cost would prevent it playing a significant role at planetary level.*
Question 3 - *Is it realistic to try to prevent the continued increase in the concentration of CO_2 in the earth's atmosphere?*
Answer - *Once again the answer is no. In the main, CO_2 emissions are the consequence of the use of fossil hydrocarbons and mankind's needs are such that it will use all the resources economically available, as they are vital to its development and to bring humanity out of poverty. What developed countries do not use will be used by China, India and other*

> nations. The official wish to halve emissions by 2050 in relation to the current level, i.e. to divide them by four in relation to the present trend, is just wishful thinking. The truth is we can do nothing about it. (Note: Since publication by Gerondeau in 2010, the Paris Agreement signed in 2015 is now targeting Net Zero emissions by 2050.)
>
> The conclusion is clear: the quantities of hydrocarbons still present underground, and nothing else, will determining the amount of CO_2 to be released in the atmosphere in the coming decades. Simple reckonings then show we have to expect the amount of CO_2 lying over our heads to about double before the end of this century.
>
> **Question 4** - *Will CO_2 emissions continue to grow indefinitely?*
> **Answer** - *As most CO_2 emissions come from the combustion of fossil hydrocarbons and as, starting with oil, these will gradually be exhausted, the answer is once again no. There will come a time when energy related emissions will hit a ceiling and start to decrease before almost disappearing at the end of the twenty first century when the reserves are exhausted.*
>
> **Question 5** - *Are we heading for disaster?*
> **Answer** - *Fortunately there is every chance this is not the case. In the distant past, the amount of CO_2 in the atmosphere has been sometimes five to twenty times what it is today, and life on earth was not harmed. No serious examination backs up the apocalyptic predictions we are fed daily about the consequences of the inevitable increase in the CO_2 concentration in our atmosphere. We are told so many obvious falsehoods regarding, for instance, the Himalayan glaciers or the effects of rising sea levels, which have no basis in scientific truths that it is hard to trust those who express them.*

What is inevitable and outside the control of the UN or any individual country irrespective of any Net Zero policies, is that the use of fossil fuels will continue unabated until all reserves that can be economically extracted are used up. The developing nations where the majority of the world's population live will consider that raising their standard of living and taking themselves out of poverty is a much higher priority for them than emissions targets or climate change. We see this scenario already taking place with China, India and other developing nations.

While accurate numbers for the reserves of fossil fuels are not known, it is estimated that at the present and anticipated extraction rates, that oil and gas reserves will be depleted by the end of the 21st century. Coal reserve estimates that until fairly recently were thought would last for another 300 years has needed a revision due to the accelerated use by China and other countries to fuel their new coal fired power generation. At these increasing consumption rates, coal reserves may now last only another 100 years. With oil and gas extraction rates declining first around the middle of the 21st century, coal will become the largest source of CO_2 emissions until its economically recoverable reserves are also exhausted. Gerondeau also discusses other options to reduce greenhouse gas emissions:

- Nuclear power is thought to continue to play only a relatively minor role in power generation globally over the next few decades due to the high initial cost, lengthy construction time and remaining widespread opposition to nuclear. However, in the longer term nuclear will have to be massively expanded as fossil fuel reserves are exhausted.

- Renewables are the chosen option by most nations intent on reducing CO_2 emissions with wind and solar the main technologies. Both these power sources are subject to intermittency, the generation of power only when the wind blows and the sun shines. Both these technologies are accompanied by significant environmental harm that is mostly hidden from the public until it directly affects them. Wind turbines generate only around 25% of their rated values due to intermittency and have to be backed up with traditional dispatchable coal or gas fired power stations.

- It should be noted that Germany has abandoned nuclear power and is transitioning to renewables. This has required the building of new gas and coal fired power stations in order to maintain supplies and stabilise the grid when the wind is not blowing and the sun not shining. As a result, this transition to renewables has actually increased Germany's CO_2 emissions from power generation. The TED Talk titled: Why renewables can't save the planet [Ref 76] provides a compelling argument against the widespread and rapid deployment of wind and solar renewables.

- Carbon capture and sequestration, also known as carbon capture and storage (CCS) of the CO_2 emissions from coal and gas fired power stations has been proposed and small pilot plants have been operated. However, this technology is extremely expensive to build and operate and is not viable without subsidies. The British Geological Survey web page: Understanding carbon capture and storage [Ref 79] provides a description. CCS is unlikely to play any significant role in emissions reduction in the foreseeable future.

Evolution of CO_2 energy emissions
Billions of tonnes per year

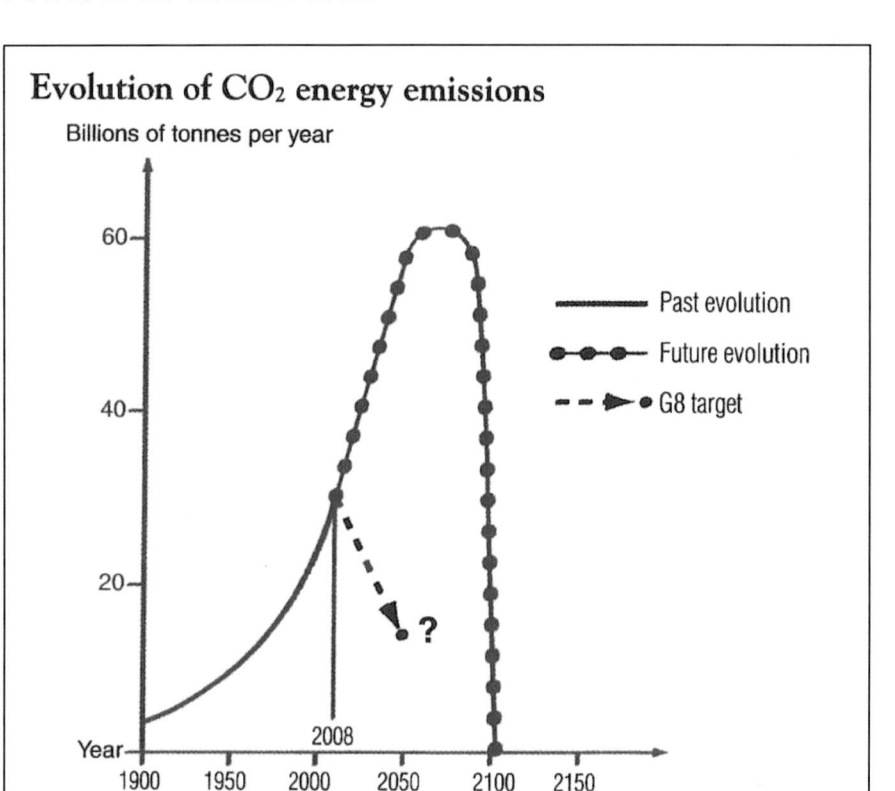

— Past evolution
●●● Future evolution
--▶● G8 target

2008

Year: 1900, 1950, 2000, 2050, 2100, 2150

Fig. A According to the reference scenario of the International Energy Agency, annual CO_2 emissions due to the use of coal, oil, and, natural gas are going to double in the middle of the century, due to emerging countries' needs.

Everything will change later on, as oil, natural gas and even coal fields will come to be depleted and emissions disappear.

The lack of realism of the target set up by the leaders of the great developed countries (G8), which is to divide by two the amount of global CO_2 emissions in 2050 is clear. To reach it, it would be necessary, not only to spend huge amounts of money in developed countries, but to stop the development of the rest of the world and leave it in poverty.

Source: Christian Gerondeau

Figure 23: Evolution of CO_2 energy emission
(Note: The G8 reduction target of 50% has since been replaced by Net Zero)

A Guide to the Climate Debate

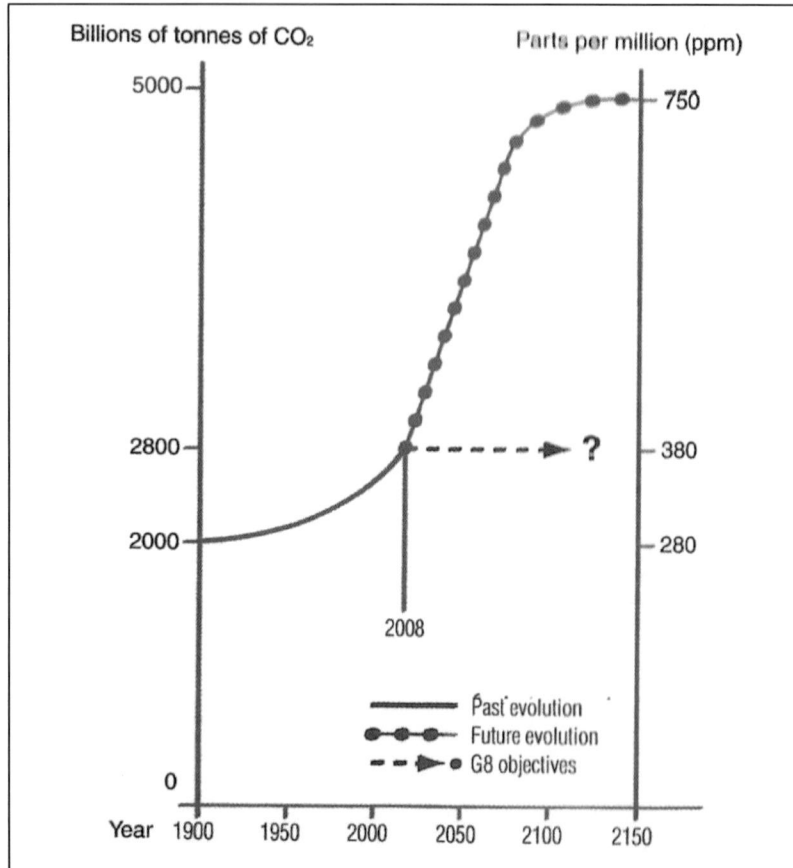

Fig. B Without the greenhouse effect, in which CO_2 undoubtably plays a small part, our planet's average temperature would be lower by perhaps as much as 30°C. Before the industrial age, the level of CO_2 in the atmosphere remained stable at around 2,000 billion tonnes, a concentration of about 280 parts per million (ppm). This concentration has increased over the past two centuries due to the burning of coal, petrol and natural gas, which has driven economic progress. Once the earth's supplies of fossil fuels have been exhausted the concentration of CO_2 in the atmosphere will stabilise before decreasing slowly over the coming centuries. The G8's objective, therefore, of maintaining the current atmospheric CO_2 concentration seems completely disconnected from reality.

Source: Christian Gerondeau

Figure 24: Presence of CO_2 in the atmosphere

See Figures 23 and 24 reproduced from Gerondeau [Ref B] with Fig A shown here as Figure 23 showing how CO_2 emissions will continue to rise until around 2050 when they will level out and peak, and then rapidly fall to zero as fossil fuel reserves are exhausted around 2100.

Fig B shown here as Figure 24 shows that the atmospheric CO_2 concentration will continue to rise in response to the increased emissions until it peaks around year 2150 at a value of around 750ppm.

What is clear is that intervention to curtail the global use of fossil fuels and their greenhouse gas emissions is not going to be effective in any meaningful way as the developing countries continue to develop and attempt to raise their populations of billions out of poverty. All the remaining economically extractable reserves of oil, gas and coal will be extracted and used by mankind until they are depleted.

These estimates by Gerondeau are based on his estimated fossil fuel reserves remaining to be burnt, so any increase in these reserves will increase the total amount of CO_2 to be emitted into the atmosphere and hence may delay and/or increase the peak CO_2 concentration depending upon usage rates.

As a check on Gerondeau's projections and estimates from 2008 we can compare these with current emissions data, 14 years later. The *'Our World in Data'* web page CO₂ emissions [Ref 80] provides annual data up to 2022. The graph *'Annual CO₂ emissions'* reproduced as Figure 25 shows that global emissions from fossil fuels and industry has continued to increase as predicted.

[Figure: Annual CO₂ emissions chart from Our World in Data, showing global emissions rising from near 0 in 1850 to approximately 37 billion t by 2022]

Source: Our World in Data

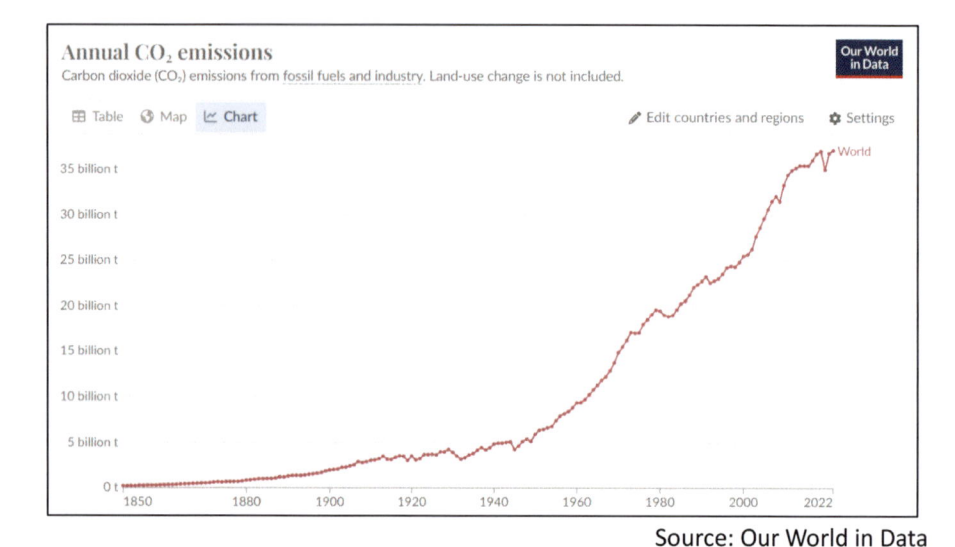

Figure 25: Global CO₂ emissions from fossil fuels and industry

Emissions of CO_2 can be seen to follow global economic activity and is unaffected by efforts by the developed countries to reduce emissions. Brief falls can be seen corresponding to the recessions during the 1930s, early 1980s, early 1990s, 2009 and the Covid epidemic during 2020. Following the Covid epidemic, emissions resumed the upward path during 2021 and reached a new high for 2022. We can therefore be confident that this trend will continue, as predicted by Gerondeau, irrespective of Net Zero policies, and until fossil fuel reserves are exhausted.

The *'Our World in Data'* graph *'Annual CO_2 emissions by world region'*, [Ref 80] reproduced as Figure 26, allows us to examine the origin of these emissions. This graph identifies that the rapid increase in emissions by China, India and the rest of Asia has more than offset the small reductions achieved by Europe and the USA. This confirms the expectation that the Net Zero policies arising from the Paris Agreement will have little effect on overall global emissions.

What this evidence shows is that any measures specifically aimed at reducing CO_2 emissions are pointless, and will not have any meaningful effect on global emissions. Any emissions saved through these measures by the developed countries, will be emitted elsewhere by others in developing countries with market supply and demand the controlling factors.

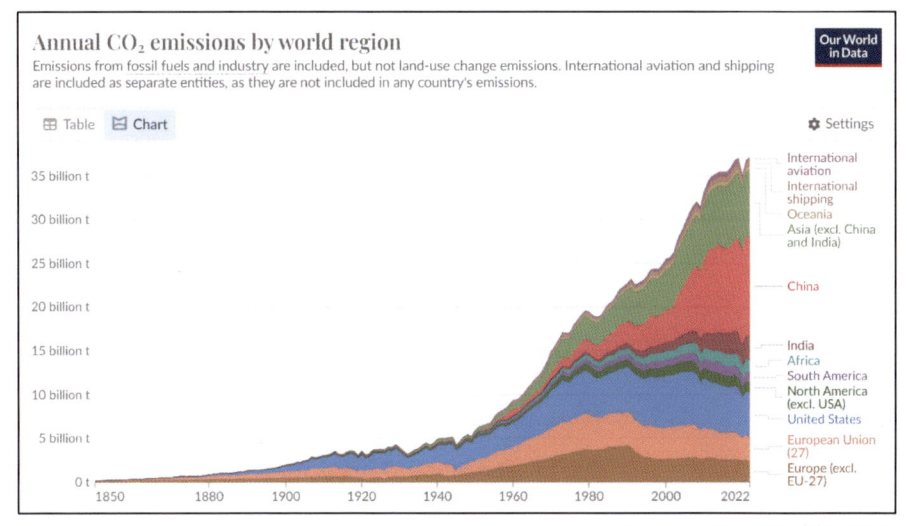

Figure 26: Annual CO₂ emissions by world region from fossil fuels and industry

Source: Our World in Data

While we can see that Net Zero policies and the attempt to eliminate CO_2 emissions by 2050, is not going to work, there are reasons, other than the attempt to control the climate, for a more gradual decarbonisation of the global economy and for conservation of remaining fossil fuel reserves:

- There is a strong argument that fossil hydrocarbons are too valuable as a chemical feedstock to be 'wasted' in direct combustion.
- As fossil fuels gradually run out during the second half of the 21st century they will become more valuable, and energy security can be expected to become a much more significant geopolitical factor if we are still fossil fuel dependent.

- As fossil fuels gradually run out during the second half of the 21st century, decarbonisation will become an economically driven necessity.

While there is no logical reason for full decarbonisation until the second half of the 21st century, there is a strong argument to replace Net Zero for a more gradual market driven decarbonisation of the global economy during the next few decades. This decarbonisation should be led by the evolving technologies, innovation and economics, not driven by subsidies and legislation.

There should also be a recognition of the likely change in public opinion regarding climate change in the not-so-distant future. The current 'climatism' driven Net Zero policies will increasingly impact the economy and our way of life, becoming more unpopular. It can only be a matter of time, for public opinion to take onboard the Global Warming deception for what it is, and to eventually turn against climate change, Net Zero, and any form of decarbonisation. This has the potential for creating political turmoil and conflict with the huge vested interests, political and financial, in the transition to renewables.

If we look to China to see how the world's largest greenhouse gas emitter plans to respond to the call for Net Zero, we can see a much more thought out and rational strategy with none of the climate change hysteria we see in the developed countries. The report by the International Energy Agency (IEA) titled: An energy sector roadmap to carbon neutrality in China [Ref 81] published 2021 provides an overview of China's strategy. The Highlights at page 18 of the report include the statements:

- *China is the world's largest emitter of greenhouse gases, at about a quarter of global
emissions. Carbon dioxide emissions from fuel combustion and industrial processes reached more than 11 gigatonnes CO_2 in 2020, of which around 90%, were from fuel combustion. Coal-fired power stations alone, including combined heat and power plants, were*

responsible for more than 45% of China's entire energy and process related emissions and 15% of global emissions in 2020.

- *In September 2020, China's president announced national aims to have CO_2 emissions peak before 2030 and to achieve carbon neutrality before 2060. CO_2 emissions per unit of GDP are targeted to fall by more than 65% between 2005 and 2030, the share of non-fossil fuels in primary energy use to around 25% and wind and solar capacity to rise to over 1 200 gigawatt (GW) in 2030 (from about 540 GW now). Growth in coal use is to be limited in the period to 2025 and phased out thereafter.*

What seems clear is that China will decide its own speed of decarbonisation that best suits its economy and its interests, based on the finite reserves of fossil fuels. The 2060 date given for decarbonisation is only given as an 'aim', not a binding target, indicating the Chinese government intend to keep this date flexible. It is also likely China will aim to maximise the benefits of cheap fossil fuels, especially coal and will only accelerate decarbonisation when the economics dictate. It is most unlikely that China will permit climatism to influence its political judgment or allow decarbonisation of their economy to adversely impact their economic development.

The strategic decision to continue the expansion of their large coal fired generation capacity will enable China to continue burning low-cost coal to generate electricity for many more decades, well into the 22nd century if needed, and likely after oil and gas reserves are exhausted.

China's strategy and decision making would seem to be based on a good understanding of the science of climate change, global fossil fuel reserves and economic reality. It may also be more than just a coincidence that China's leaders come mostly from science and engineering backgrounds, unlike the UK where very few politicians have such technical backgrounds.

We see a similar situation with India, the 3rd largest global emitter of greenhouse gases. India has declared a nominal target date of 2070 to achieve Net Zero and has also opted out of the Paris Agreement. India also continues to expand its coal fired generating capacity as it expands its economy and will be heavily dependent on fossil fuels for many more decades.

A Guide to the Climate Debate

Q12. What are the alternatives to Net Zero?

As we have seen when investigating key question 11, Net Zero policies adopted by mainly western developed countries will be ineffective in preventing the continued increase in greenhouse gas emissions from the burning of fossil fuels. While there are still billions of humans on the planet living in poverty, the developing countries will prioritise their own development over the reduction of emissions. These greenhouse gas emissions will only reduce when all economically recoverable reserves of oil, gas and coal have been extracted and used, estimated to be around year 2100. At this time Net Zero will occur by default.

By implication, the introduction of Net Zero policies some 50 years ahead of the default Net Zero by some of the developed countries is pointless as they are guaranteed to fail. These ideologically driven policies simply divert resources and attention from the more important needs in the world today and will likely bankrupt the economies of the countries attempting to meet Net Zero targets.

What about the science?

When investigating our key questions 1 to 8 we have examined the science behind climate change. We find conclusive evidence that the warming and cooling cycles as observed from the past have been driven by cycles in the radiated energy from the sun, known as the total solar irradiance (TSI). Changes in atmospheric CO_2 concentration have been shown to be a response to these warming and cooling cycles, driven mainly by degassing or absorption of CO_2 by the oceans. The changes in atmospheric CO_2 concentration typically lagging temperature by several hundred years.

The change in CO_2 concentration during these climate cycles almost certainly will have contributed to some climate forcing but the degree of forcing has never been quantified. However, the switching, the changes from warming to cooling and back to warming have been shown to be controlled by the solar TIR cycles. There is no evidence to suggest that the warming phase observed in the climate today is significantly different from what has been observed from the past. Also, the global climate was warmer during the Medieval warm period, less than 1,000 years ago than it is today.

The climate change hypothesis supported by the Warmists including the IPCC and most western governments that climate change is the result of increased CO_2 emissions caused by human activities, known as Anthropogenic climate change, is not supported by the evidence. After over 40 years and the spending of billions of dollars on research into climate change, there is still no proof of this theory. There is though strong evidence to the contrary as outlined in the examination of key questions 1 to 8 above.

> If we could magically and instantly, reduce all man-made CO_2 emission globally to zero, in effect an instant Net Zero, it is unlikely to make any significant difference to the climate. As a result, the implementation of Net Zero policies by the UK and other industrialised countries to eliminate man-made CO_2 emissions even if these policies could be made to work, will be ineffective in controlling the climate.

What about the climate?

With the evidence showing that Net Zero policies will be ineffective in limiting global greenhouse gas emissions, we anticipate the atmospheric CO_2 concentration will continue to rise until it reaches a peak of around 750ppm by the year 2150 as per Figure 24. After year 2100 with CO_2 emissions from the burning of fossil fuels falling rapidly, the atmospheric CO_2 concentration will be expected to slowly reduce over the following decades as the natural carbon cycle recycles the atmospheric CO_2 into the biosphere via plant photosynthesis and by absorption into the ocean.

A Guide to the Climate Debate

During this period of high atmospheric CO_2 concentration and likely a warmer climate, plant growth rates will be higher than today. This higher growth rate is expected to increase the rate of transferring carbon from the atmosphere to the biosphere so further speeding the reduction of atmospheric CO_2.

We discovered the effects on plant growth from elevated CO_2 concentrations at key question 3, see Figure 7. Commercial greenhouse growers are using elevated CO_2 concentrations of typically up to 1,000ppm to provide an enhanced atmosphere to increase crop yields and to prevent CO_2 depletion as the plants grow and use up the available CO_2. Since CO_2 is harmless, the projected CO_2 levels peaking at around 750ppm by 2150 should not in itself be a cause of any health concerns. 750ppm is still a very small amount, just 0.075% of the atmosphere. Note that the CO_2 concentration in crowded meeting rooms and other closed indoor spaces can often exceed 3,000ppm simply due to our breathing.

While there is nothing, we can do to control the climate or atmospheric CO_2 concentration, we should look to the past in order to understand how the climate could change over the rest of the 21st century and beyond. If we refer to Figure 10, we have to go back 4 to 5 million years to find a time when temperatures have been 4°C higher than the recent pre-industrial times. For most of this period global temperatures were well below the pre-industrial reference temperature.

Given the logarithmic relationship that exists between increased CO_2 and a forced increase in temperature due to the greenhouse effect, a doubling of the current concentration of CO_2 in the atmosphere will in itself only cause a temperature rise of around 1°C. In fact, the climate records indicate the earth faces a greater risk and danger from global cooling than warming over the next decades and centuries.

We also examined climate prediction at key question 9 including the deterministic GCM computer models used by the IPCC. We have also seen how these GCMs have been responsible for some of the alarmist climate messages over the last decades, and that these climate projections have consistently proved to be inaccurate, being no more that climate virtual reality.

An alternative methodology to the deterministic GCMs used by the IPCC is statistical climate modelling, but unfortunately is not used by the IPCC. This is a well understood scientific alternative and uses data from past climate change to construct projections of future climate. Several of these models have independently arrived at similar conclusions by forecasting cooling for the first decades of the 21st century followed by a warming. Depending upon the climate model you choose, you can generate outputs that predict either warming or cooling. Green *et al.* [Ref 53] found that the IPCC model errors were seven times greater than the alternative statistical 'benchmarking' method.

> The bottom line on climate prediction is that scientists are a long way from being able to predict the future climate despite what the IPCC would like you to believe.

Climatism

A major barrier today to any change in direction with climate policy in western developed countries is the embedded new religion known as climatism, already encountered in our review of evidence. In his book titled **'Climate Change Isn't Everything'** [Ref C] published in 2023, Mike Hulme provides an alternative view on climate policy, climatism, and some valid reasons for a change in direction.

Hulme is a staunch Warmist climate scientist and was involved in the periphery of the Hockey Stick deception by Warmists described by Montford

> Hulme defines climatism as, *"a settled pattern of ideas, beliefs and values which hold that the dominant explanation of social, political and ecological phenomena is a changing climate"*. The underlying belief of climatism is that preventing the climate from getting worse will prevent everything else from getting worse.

A Guide to the Climate Debate

in the book '**Hiding the decline**' [Ref D], discussed at key question 9. It was therefore quite unexpected to find a book with such a title written by Hulme. However, in his book Hulme stays away from the science of climate change and instead focuses on the dangers of climatism. He does though make several Warmist statements throughout the book, possibly to reaffirm his credentials with the Warmist community. He goes on to provide a description of climatism as an ideology where everything that is going wrong in the natural world and the political world and all its ills can be shown to have climate change as their underlying cause. Hulme explains that climatism is now a fully fledged ideology and in the process of becoming a fully fledged dogma.

It leads to climatic events that previously would have been considered 'Acts of God', have now been replaced by 'Act of Man' and the percentage of blame that is due to man-made climate change has been apportioned. For instance, when Typhoon Hagibis hit Japan in October 2019, 40% of the ensuing $10 billion worth of damage was attributed to climate change. The implication is that for victims of 'Act of God' it is tough luck, but now victims of 'Acts of Man' will be demanding to be compensated. This opens the possibility of specific countries or companies who have emitted a certain amount of CO_2 being held responsible for specific deaths or damage. This is becoming big brother on a global scale.

Hulme is surprisingly critical of the IPCC climate models believing that the projections from the IPCC sixth assessment report (AR6) in 2023 predict the climate will get too hot, too fast. By implication he is saying the IPCC is misleading policy makers. It is notable that Hulme uses the word 'predict' when referring to the IPCC 'projections', a topic discussed above at key question 9, General Circulation Models. He even recognises that the science and the ideology have become self-reinforcing, and as he states "*a close and interdependent relationship*".

Hulme highlights the IPCC Assessment reports as only identifying the negative impacts of climate change while ignoring the positive impacts. Since these reports are used by government policy makers, he acknowledges the reports are misleading. As an example, he points to the BBC website for children that was providing information on the benefits as well as the negative impacts of climate change. However, after complaints by advocacy groups, the BBC website now only provides information on the negative impacts of climate change. When it comes to uncertainties with the science, Hulme quotes a prominent American climate scientist, the late Stephen Schneider, quote;

> *"On the one hand, as scientists we are ethically bound to the scientific method, in effect promising to tell the truth, the whole truth, and nothing but the truth but – which means that we must include all the doubts, the caveats, the ifs, ands, and buts. On the other hand... to reduce the risk of potential disastrous climatic change...entails getting loads of media coverage. So, we have to offer up scary scenarios, make simplified, dramatic statements, and make little mention of any doubts we might have... Each of us has to decide what the right balance is between being effective and being honest. I hope that means both."*

The Schneider quotation describes what is sometimes referred to as the 'Noble Lie' and has been discussed earlier at key question 10 under the section **'A noble cause'**. As we have already discovered, some scientists have pushed this balance firmly towards climatism.

Hulme explores why climatism is so alluring. For some climatism brings together the seemingly disparate issues of protecting humanity from a destabilised climate system with what is believed to be an unjust capitalist economic system into a single coherent narrative. Hence the obvious attraction to left wing activist groups and left-wing politicians. He quotes the American campaigner Michael Shellenburger on describing the allure of climatism for many,

A Guide to the Climate Debate

> *"It is powerful because it has emerged as the alternative religion for supposedly secular people, providing many of the same psychological benefits as traditional faith. It offers a purpose – to save the world from climate change – and a story that casts the alarmists as heroes. And it provides a way for them to find meaning in their lives – while retaining the illusion that they are people of science and reason. Not superstition and fantasy"*

Hulme then moves on to the dangers of climatism and starts by describing the disastrous EU policy on biofuels, primarily for use as an additive to petrol, by diluting it with ethanol with the aim of reducing CO_2 emissions. The CO_2 emissions from the ethanol percentage not counting, as these emissions are considered 'sustainable'. This policy is based on the belief that climate change is everything and ignores any negative impacts of this policy.

As a result, Sumatra in Indonesia has lost 50% of its tropical rainforest in the last 35 years, driven largely by the growth in palm oil plantations used for these bio-fuels. This has had devastating environmental impacts plus devastating social consequences for the indigenous people who lived there. Subsequent analysis has since shown that the use of this ethanol bio-fuel has not actually saved any CO_2 emissions, and may have actually increased emissions overall.

This is a similar example of greenwashing as described in the Preamble section, of the burning of hardwood from North American old growth forests at the UK Drax power station. All justified by the narrow vision of climatism to cut CO_2 emissions, to affect the climate in 50 years' time, and irrespective of the interim costs, financial and environmental.

Hulme deals with scarcity, the practice of imposing deadlines by which date, action will be too late and encouraging a climate state of emergency. Imposing these deadlines causes psychological effects and inhibits human cognitive capacity to imagine a future beyond the deadline. It has resulted in surveys showing large proportions of young people aged 16 to 25 feeling sad, afraid, anxious and many thinking humanity is doomed. We now have a generation of young people growing up in an environment of fear and hopelessness for the future.

Hulme discusses the danger of depoliticising and explains how climatism depoliticises climate change and makes the comparison with the response during 2020 to Covid 19. An issue can be said to have become depoliticised when for example a society or government claims *'the science leaves no alternatives'* or to be *'following the science'*, or *'listening to the scientists'*.

Aided by the shortness of time, the politics of climate change is reduced to simply delivering the technical and economic mechanisms of securing Net Zero. The values, choices and trade-offs remain protected from public view and scrutiny. We see this already taking place in the UK with no more discussion of the why, only the how. The election of the Labour government in July 2024 has accelerated this push for Net Zero at any cost.

Depoliticising climate change has effectively closed off the possibility of advocating or even debating other policy goals. This leads to the related risk of endangering democratic values and the delegitimization of dissent. Hulme provides examples of how this is already taking place including the delegitimization of respected scientists simply for questioning the transition to renewables and advocating greater use of nuclear power.

A Guide to the Climate Debate

Figure 27: UN Sustainable Development Goals agreed 2015

Source: United Nations

Hulme then moves on to consider what should be the global priorities in addition to the climate. In 2015 the United Nations agreed the seventeen Sustainable Development Goals (SDGs) illustrated at Figure 27.

On the UN Sustainable Development Goals web page [Ref 82] it states,

"The 2030 Agenda for Sustainable Development, adopted by all United Nations Member States in 2015, provides a shared blueprint for peace and prosperity for people and the planet, now and into the future. At its heart are

A Guide to the Climate Debate

the 17 Sustainable Development Goals (SDGs), which are an urgent call for action by all countries - developed and developing - in a global partnership. They recognize that ending poverty and other deprivations must go hand-in-hand with strategies that improve health and education, reduce inequality, and spur economic growth – all while tackling climate change and working to preserve our oceans and forests".

The overall objectives of these 17 SDGs then are aimed at ending poverty and other deprivations alongside strategies that improve health and education, reduce inequality, and spur growth. All while tackling climate change and working to preserve our oceans and forests.

Then a few months later in December 2015, the UN completed the Paris Agreement, signed by the same nations. This, agreement now effectively overrides the SDGs with the Paris Agreement goals of emission reduction and global temperature control becoming an overriding priority. Although the Paris Agreement recognises the context of sustainable development, the goal of securing global temperature within a certain numerical range then took precedence over a broader set of SDG welfare ambitions.

Hulme uses as an example of the narrowness of climatism thinking with the case of 2.5 billion people in the poorer nations reliant upon cooking on stoves or fires fuelled by kerosene, coal, animal dung or other forms of biomass. The World Health Organisation estimates that about 3.8 million premature deaths are caused annually linked to the household air pollution created by these cooking arrangements. SDG 7 relates to this third world problem to ensure access to affordable, sustainable and reliable energy for all.

One of the most affordable, clean and scalable solutions for replacing this harmful practice is with liquid petroleum gas (LPG) cooking stoves. However, the fixation on one goal by the Paris Agreement is preventing this from being provided as LPG is a fossil fuel and a source of CO_2 emissions.

This therefore brings us to the title of Hulme's book **'Climate change isn't everything'** in that there are many problems affecting the world as shown by the 17 SDGs with climate change just one of them. He argues that a future world where global temperature rise exceeds 2°C but with political stability and ecological integrity for example, could be a better place than one where the temperature was limited to 1.5°C at all costs.

He reaffirms that climate change isn't everything, and that there needs to be trade-offs between the SDGs and Net Zero, in effect climate pragmatism. In response to the alarmist and existential risk warnings of climate change being promulgated he tempers this with the statement, *"There is no good scientific or historical evidence that climate change will lead to human extinction or the collapse of human civilisation"*.

The surprising aspect of this book is that Hulme is so critical of many aspects of climate change and the politics of Net Zero. The only areas he carefully avoids are the science itself and the fact that Net Zero will not work and cannot ever be made to work. Had he strayed into these areas he would no doubt have fallen foul of Warmist establishment, labelled as a Denier, and his academic career would be over. It is suspected that Hulme is a closet Sceptic.

We need a Plan B

What has become apparent is that we need a Plan B, globally and for the UK for responding to climate change since:

a) We are not able to predict the climate, and;
b) Climate is controlled by the changes in the energy received from the sun, and not by man-made CO_2 emissions, such that;
c) Net Zero emission policies (Plan A) can never control the climate, and;
d) CO_2 emissions saved by developing nations will be emitted by developing nations;
e) Therefore, Net Zero targets can never be achieved in today's real world with the developing nations opting out, and will simply bankrupt any nations that attempt it.

Since Plan A (Net Zero by 2050) will not work and cannot be made to work, irrespective of whether anthropogenic climate change is real or not, we therefore need a Plan B. The main features of a Plan B should logically include recognition that:

a) There will be an eventual default Net Zero around year 2100 when global reserves of oil, gas and coal have been exhausted.
b) We are not able to predict or control the climate.
c) Natural climate change over the next few decades is unknown and could be a continuation of the current warming trend, or a change to a cooling trend, or an acceleration in either a warming or cooling direction at any time.
d) Natural climate change presents a future risk to the world although there is no evidence of an impending end of humanity or of human civilisation.
e) Efforts to control the climate should be replaced with mitigation, aimed at anticipating and managing the effects of climate change locally.
f) CO_2 is not a pollutant or harmful and is essential for life on earth. CO_2 emissions from human activities do not present any danger and do not need to be controlled.
g) The current focus on CO_2 emissions should move to real environmental issues including pollution of the air, land, rivers and ocean, deforestation and declining biodiversity.
h) Renewed priority should be given to the United Nations SDGs discussed above.
i) Decarbonisation of the global economy should be encouraged but at a slower natural rate driven by economics, technology and the environment, to facilitate the default Net Zero around year 2100.

Great spirits have always found violent opposition from mediocrities. The latter cannot understand it when a man does not thoughtlessly submit to hereditary prejudices but honestly and courageously uses his intelligence.

(Albert Einstein 1879-1955)

A Guide to the Climate Debate

Findings and Conclusions

The Science
The findings at key questions 1 to 8 can be summarised by the following:

- The earth is warming with the global average currently around 1.2°C warmer than during the pre-industrial period. What is also evident is that alongside this warming, changes to the climate are not uniform around the world with some location's wetter and some drier. There is also some doubt as to the accuracy of the global temperature rise due to uncertainties with the UHI effect. While the earth is shown to be warming, climate history tells us the current warming trend is not unusual. It was warmer during the Medieval Warm Period, less than 1,000 years ago.

- The evidence shows that extreme weather has increased overall although the changes are not uniform around the globe. Whereas heatwaves and flooding have seen an increase, there is mixed data on the trend for hurricanes and an overall decline for wildfires.

- We established that the natural carbon cycle is central to life on earth and its climate. The ocean with its dominance in both the carbon cycle and the heat cycle effectively controls the atmospheric CO_2 concentration, and the earth's climate. Despite significant scientific advances there are still large unknowns and uncertainties with the natural carbon cycle.

- The rising atmospheric CO_2 concentration is caused by a combination of human activities and natural effects but the absolute contributions cannot be quantified due to the many uncertainties. The uncertainties associated with emissions from volcanoes, especially undersea volcanoes make any meaningful estimates impossible.

- While the theory of the greenhouse effect is well understood, the mechanisms by which the earth's atmosphere responds to the greenhouse effect is much more complex and is only partially understood. The greenhouse effect of CO_2 is relatively small compared to that of water vapour, the main greenhouse gas. Additionally, the greenhouse effect of CO_2 is limited due to the declining logarithmic response to increases in CO_2 concentration.

- The sun is the dominant global heat source and its energy drives the earth's climate system. The sun's Total Solar Irradiance (TSI) is not constant and its these cycles of radiation intensity, and not atmospheric CO_2 concentration, that have been shown to control the earth's climatic variations. There are also significant uncertainties with the global heat cycle including the effects of cloud cover, effects of galactic cosmic rays and uncertainties with the variability of the solar wavelength spectrum.

- The evidence shows that global warming and cooling cycles have been driven naturally by the cycles of Total Solar Irradiance (TSI). There is a proven correlation between solar activity and global temperature, but no correlation between the world's hydrocarbon use and global temperature. **This evidence is inconsistent with and effectively disproves the IPCC climate change hypothesis.**

- There is historical evidence that global temperature cycles precede changes in atmospheric CO_2 concentration. This evidence shows that the warming or temperature rising phases have preceded the rises in atmospheric CO_2 concentration by several hundred years. This delayed rise in atmospheric CO_2 concentration is consistent with outgassing from the ocean as it warms. **This evidence is inconsistent with and effectively disproves the IPCC climate change hypothesis.** The IPCC climate change hypothesis dictate that rising CO_2 should precede and is the cause of an increase in global temperature. If this was true, the positive feedback effect would lead to a thermal

A Guide to the Climate Debate

runaway. This has never occurred. This thermal runaway effect, also referred to as a 'tipping point' was one of the earlier alarmist predictions, later shown to be false.

- Global sea level is rising having risen between 210mm and 240mm since 1880. This rise is due to thermal expansion of the ocean due to warming and the melting of land-based ice sheets and glaciers. Although there has been a slight increase in the rate of rise in the 20th century there is no evidence that the rate is increasing significantly or that human activities are having a measurable effect. Given the long time constant for the ocean of around 1,000 years it is anticipated that the current rise will continue at around the present rates. However, sea level rise does present a risk for some locations around the world where the land is sinking. The rise in sea level around the UK by year 2100 assuming a 3°C global temperature rise, is projected by the IPCC to be between 40cm and 70cm, depending upon location.

- The continued high atmospheric CO_2 concentration will continue to have a slight acidification effect on the ocean. Sea water is alkaline and always has been, and currently the pH is typically 7.5 at depth and 8.5 in surface waters although it is locally variable. (Note: a pH of 7 is neutral, greater than 7 alkaline, and less than 7 is acid) The projection for year 2100 is for surface seawater pH to change to around 7.8. There is no evidence that this slight acidification will have any significant effect on marine life. However, there is a recognition that marine life could be affected locally, in tropical areas due to surface sea water warming.

The Politics

The findings at key questions 9 to 12 can be summarised by the following:

- What started during the 1970s as a theory by a minority of scientists that a rising atmospheric CO_2 concentration could create global warming, has grown over four decades into a widespread belief in anthropogenic climate change. Also known as the climate change hypothesis, this theory has now become an ideology, akin to a religion for many and is known as climatism. The change from a belief to the ideology of climatism has been characterised by alarmist predictions and propaganda for the last 40 years. Most predictions have been made by the media and environmental groups, often quoting claimed climate experts. Most of these alarmist reports when examined have been found to be factually wrong, misleading and simply did not materialise. However, the propaganda has facilitated the development of climatism.

- The development of climatism has been made possible by scientists straying from their strict scientific discipline, convinced of the morale righteousness of the cause. Gradually the cause was taken up by many well-meaning and influential individuals, but also environmental groups, renewable energy companies and politicians eager for 'green' votes. A key factor in the development of climatism has been the lack of understanding of climate science by most of the population and a trust and belief in the claimed experts. In the UK these experts include the likes of Sir David Attenborough, the media including the BBC and the 'establishment' including King Charles. Many being well-intentioned but misinformed of critical aspects of the science.

- During 1988 we saw the creation by the United Nations of the Intergovernmental Panel on Climate Change (IPCC) and then its subsequent growth to a quasi-scientific but mainly political body that influences governments and their billions of dollars of green investment. The IPCC is entirely dependent on the continuation of the widespread belief in anthropogenic climate change and the continuation of an alarmist agenda for its continued existence. The IPCC therefore has no interest in consideration of alternative scientific evidence challenging the status quo or alternatives to Net Zero policies.

A Guide to the Climate Debate

- Climate change has been enthusiastically adopted by those extremist political and environmental groups where the politics of fighting climate change are integral with opposing capitalism, globalisation and control of society. Such groups including Greenpeace and Friends of the Earth are now fully integrated into the workings of the IPCC. Climatism has also become associated with the advantages it confers on capitalism, globalisation and control of society, in other words, the establishment / elite. Hence the lobbying from big business (capitalism) that aims to maintain the global warming momentum at the expense of their useful fools (Sir David Attenborough, the media, King Charles, Green Peace, Friends of the Earth, Just Stop Oil, etc.). One has only to follow the money.

- We find the climate models used by the IPCC and others known as General Circulation Models (GCMs) are only capable of generating 'projections' to aid research. The limitations of climate science and the associated uncertainties prevent any meaningful climate forecasts or predictions. It is simply not possible to predict the climate years or decades ahead as is often claimed, any more than we can forecast the weather more than a few days ahead. However, the GCMs are being misused with the uncertainties conveniently ignored. The 'projections' from these GCMs are often the source of the alarmist climate 'predictions' appearing in the popular media.

- There is clear evidence of manipulation of the science to propagate the now widespread belief in anthropogenic climate change. This manipulation of the science was exposed in 2001 with the infamous 'Hockey Stick' deception where Climate Research Unit (CRU) scientists and the IPCC presented falsified data supporting their argument for anthropogenic climate change. A key objective of the Hockey Stick deception was the erasing of the 'Medieval Warm Period' from the history books, a time when conditions on earth around 1,000 years ago were warmer than today. Removal of the Medieval Warm Period enabled the Warmist argument that the current global warming was unprecedented. The subsequent leak of

CRU emails during 2009 becoming known as 'Climategate' confirmed the 'hockey stick' deception and exposed widespread data manipulation, manipulation of the peer review process and attempts by the CRU to silence dissent. These revelations were subsequently the subject of an 'establishment' cover up with the official investigations considered a 'whitewash'.

- In many countries including the UK and the USA, climate science dissent has been effectively eliminated. This has been achieved gradually over the years by control of the funding, control of the scientific journals and by the 'establishment'. It would now be either a brave move or career suicide for any scientist in the UK today to present evidence challenging the climate change hypothesis.

- In the UK, the Royal Society, a bastion of science and the UK 'Establishment' has betrayed all that science stands for including the *'scientific method'* by supporting the idea that scientific truths should be established by a political process of consensus, and in effect suppressing dissent. The Royal Society claims that anthropogenic climate change has been proved through the use of climate modelling, so not through the *'scientific method'*. The position taken by the Royal Society demonstrates that the 17th century experience by Galileo over his claim that the earth revolved around the sun has in effect been repeated in the 21st century with anthropogenic climate change.

- After the spending of billions of dollars on climate research for over four decades, there is still no scientific proof of anthropogenic climate change. There is however, proof that climate change is controlled by the cycles in the energy received from the sun, the solar TIR, evidence that the IPCC prefers to ignore.

- The Paris Agreement signed in 2015 introduced the target of limiting global warming to 2°C, but preferably 1.5°C to be achieved by reducing man-made CO_2 emissions globally to Net Zero by 2050, all in the belief that this will enable control of the climate. Many of the developed countries are now adopting Net Zero policies including the UK. However, major emitters including China, India and the developing countries have opted out of the 2050 target date, instead they will decarbonise to their own agenda to suit their own economic interests.

- The evidence shows that global CO_2 emissions continue to climb unabated and in line with earlier projections. Any emissions savings achieved by the developed countries are emitted by developing countries. As heavy industries such as steel and cement making are curtailed in the developed countries, these industries are expanded in China and the developing countries.

- It is totally unrealistic to expect the developing countries to reduce their already low per capita CO_2 emissions while there are still billions of people living in poverty. It is also totally unrealistic to expect humanity to leave any economically recoverable fossil fuel left in the ground. The developing countries will continue to burn fossil fuels until all has been extracted as their economies develop, simply because they have no other option.

- Through the continued use of fossil fuels, the resulting CO_2 emissions are predicted to continue to climb until the second half of the 21st century when they will peak, then fall rapidly as fossil fuel reserves are exhausted, at approximately the end of the 21st century. The atmospheric CO_2 concentration will also continue to rise and is likely to peak at around 750ppm at the end of the 21st century.

- The Paris Agreement and subsequent individual country Net Zero policies therefore have zero possibility of reducing CO_2 emissions globally to zero by 2050. Net Zero will only be achieved after year 2100, and then by default when all economically recoverable reserves of oil, gas and coal have been extracted and burnt.

- Contrary to the IPCC climate change hypothesis, the scientific evidence confirms that the changes in the climate are determined by the cycles in solar TSI, and not the level of CO_2 in the atmosphere. Therefore, even if global man-made CO_2 emissions could be instantly reduced to zero it would not have any significant effect on the climate. The continued pursuit of Net Zero policies by individual countries will simply serve to destroy their economies and way of life while having zero effect on the climate.

- Since it is not possible to control or even predict the climate, the Paris Agreement and Net Zero policy should be abandoned. With or without Net Zero policies, emissions from the burning of fossil fuels will peak in the second half of the 21st century and atmospheric CO_2 will continue to rise to a concentration of around 750ppm around year 2100. Contrary to the alarmist warnings, the effect of an almost doubling of the CO_2 concentration is estimated to raise global temperature by only around 1°C. Fortunately, CO_2 is not a pollutant or harmful and is essential for life on earth. The reality is that CO_2 emissions from natural causes or human activities do not present any danger to the world and do not need to be controlled.

- There needs to be a recognition that natural climate change is taking place, always has done and will continue to do so into the future, presenting risks to various locations around to the world at various times. However, there is currently no evidence scientific or historical of an impending climate emergency, the end of humanity or human civilisation. In addition to the risks or negative aspects to a warming climate, there are also considerable benefits. A benefit already being realised of an increased atmospheric CO_2 concentration and a warming climate is increased plant growth and crop yields.

- Natural climate change cannot be predicted and could be a continuation of the current warming trend, or a change to a cooling trend, or an acceleration in either a warming or cooling direction.

A Guide to the Climate Debate

Climate change mitigation should be aimed at anticipating, adapting to and managing the effects of climate change locally.

- Today we see 'Net Zero at any cost' around the developed world, especially in the UK along with the adverse unintended consequences associated with the deployment of renewables and the widespread 'greenwashing'. The destruction of tropical rainforests in Asia to grow palm oil for bio fuel and the felling of North American hardwood tree for burning in the Drax power station are unfortunate and destructive examples. In both these cases, investigation has shown the overall CO_2 emissions have actually increased, not decreased, but as a bio-fuel both wood pellets and ethanol are considered 'sustainable'.

- This type of greenwashing has effects closer to home with widespread environmental harms inflicted on the UK countryside and on the people living there through the deployment of subsidised wind and solar renewable energy and new electricity transmission lines. This environmental harm is a direct consequence of entrenched climatism and unachievable Net Zero policies, enforced on mainly rural communities using the Planning Inspectorate to override local democracy.

- In 2015 the UN agreed the Sustainable Development Goals (SDGs) for the world. These SDGs identify the priorities, mostly affecting the developing countries, aimed at ending poverty and other deprivations alongside strategies that improve health and education, reduce inequality, spur growth and for protection of the environment. Unfortunately, the Paris Agreement signed a few months later and by the same countries has had the effect of overriding the agreed SDG priorities with Net Zero elevated as the overriding priority. These SDGs show that climate change isn't everything and addressing the SDGs should be elevated as the top priority for the global community.

A Guide to the Climate Debate

- There is an urgent need for an alternative plan for responding to climate change, withdrawal from the Paris Agreement and Net Zero policies. Abandoning this Plan A and replacing it with a new Plan B focused around the United Nations SDGs will permit a renewed focus on fighting poverty and environmental issues. Abandoning Net Zero would offer a more pragmatic way to address the world's most pressing needs including mitigating the effects of climate change.

- Importantly, a Plan B would also allow the redirection of huge financial resources currently being wasted on the futile attempt to control the climate to the many worthy causes. In the UK, abandoning unachievable Net Zero policies would release huge financial resources for tackling many of the county's many pressing social and environmental needs.

Some final thoughts

This review of the science and politics of climate change was triggered in early 2024 following a viewing of the documentary Climate the Movie [Ref 1]. A documentary that challenged this author's then belief in anthropogenic climate change and led to the subsequent formulation of 12 key questions that needed addressing. The subsequent investigation was found to be much more extensive than originally envisaged leading to the writing of this guide to the climate debate.

Now after months of research and investigation, these key questions have been addressed. The overall conclusion is that the evidence examined fully supports the arguments made in **Climate the Movie** and has changed this author's belief in anthropogenic climate change. Climate the Movie is highly recommended viewing for anyone interested in climate change and establishing the truth.

The evidence reviewed for this guide to climate change is readily available in the public domain, and when analysed comprehensively dismantles and

A Guide to the Climate Debate

disproves the IPCC climate change hypothesis. Unfortunately, climatism is now so firmly entrenched in many western democracies including the UK that it could be decades before common sense prevails, public opinion changes and Net Zero policies are abandoned.

These findings from this investigation into climate change can be summarised by the following:

---
a) We are not able to predict the climate, and;
b) Climate is controlled by the changes in the energy received from the sun, and not by man-made CO_2 emissions, such that;
c) Net Zero emission policies can never control the climate, and;
d) CO_2 emissions saved by developing nations will be emitted by the developing nations;
e) Policies targeting Net Zero by 2050 can never succeed in today's real world with the developing nations opting out, and will simply bankrupt any nations that attempt it.

Since Plan A (Net Zero by 2050) will not work and can never work, we therefore need a Plan B. The main features of a Plan B should include recognition that:

A Guide to the Climate Debate

a) There will be an eventual default Net Zero around year 2100 when global reserves of oil, gas and coal have been exhausted.
b) We are not able to predict the climate.
c) We are not able to control the climate.
d) Natural climate change over the next few decades is unknown and could be a continuation of the current warming trend, or a change to a cooling trend.
e) Natural climate change presents a future risk to the world although there is no evidence of an impending climate emergency or threat to humanity.
f) The futile attempts to control the climate should be replaced with mitigation, aimed at anticipating and managing the effects of climate change locally.
g) CO_2 is not a pollutant or harmful and is essential for life on earth. CO_2 emissions from human activities do not present any danger and do not need to be controlled.
h) The current focus on CO_2 emissions should move to real environmental issues including pollution of the air, land, rivers and ocean, deforestation and declining biodiversity.
i) Renewed priority should be given to the United Nations SDGs.
j) Continued decarbonisation of the global economy should be encouraged but at a slower rate driven by economics, technology and the environment, to prepare for the default Net Zero around year 2100.

The pessimist complains about the wind; the optimist expects it to change; the realist adjusts the sails.
(William Arthur Ward (1921-1994)

A Guide to the Climate Debate

Appendices

1 - Glossary

This section provides additional information on the abbreviations and technical terms used in this report.

Name / Abbreviation	Definition / Description
AMOC	Atlantic Meridional Overturning Circulation. The AMOC is a system of ocean currents that circulates water within the Atlantic Ocean, bringing warm water north and cold water south. The global conveyor belt circulates cool subsurface water and warm surface water throughout the world.
Anthropogenic climate change	Climate change caused by human activities
AGW	Anthropogenic Global Warming
AR4	IPCC Fourth Assessment Report
AR5	IPCC Fifth Assessment Report
AR6	IPCC Sixth Assessment Report
Biosphere	The biosphere is made up of the parts of Earth where life exists. The biosphere extends from the deepest root systems of trees, to the dark environment of ocean trenches, to lush rainforests and high mountaintops. Scientists describe Earth in terms of spheres.
CCS	Carbon Capture and Storage, or Carbon Capture and Sequestration
Climate Audit	A Sceptic blog website run by Steve McIntyre, a climate scientist; https://climateaudit.org/
Climate Research	A peer reviewed scientific journal published by the Inter-Research Science Centre. - https://www.int-res.com/journals/cr/cr-home/

A Guide to the Climate Debate

Climatism	The ideology of Climatism can be described as the belief in Anthropogenic climate change and its doomsday effects for the world without rational thought or full examination of the scientific evidence. It is part of a recent trend in eco-religion in western society where science is ignored but scientific words are used as cover as part of the belief system. Believers consider anyone sceptical of global warming to be an enemy of the environment when in fact many Sceptics are passionate about protecting the environment.
CO_2	Carbon Dioxide is a chemical compound made up of molecules that each have one carbon atom and two oxygen atoms and exists as a colourless gas at room temperature. It can be liquified by pressurising to 1,071 psi @ 31°C. If allowed to expand the liquid cools and partially freezes to a snow like solid known as dry ice at -78.5°C under normal atmospheric pressure, then subsequently vaporises.
CRU	Climatic Research Unit, of the University of East Anglia
Denier	A derogatory term with overtones of 'Holocaust denier' often used to describe person that disagrees with or challenges the any aspect of Anthropogenic climate change.
et al.	Abbreviation of the Latin term "et alia" which means "and others".
FAR	IPCC First Assessment Report
FOI	The UK Freedom of Information Act 2000 provides public access to information held by public bodies.
GCM	General Circulation Model – Computer models used by the IPCC for simulating the response of the global climate system
GCRs	Galactic cosmic rays
Greenwashing	The act or practice of making a product, policy or activity appear to be more environmentally friendly or less environmentally damaging than it really is.
GRL	Geophysical Research Letters, is a peer reviewed scientific journal of geoscience published by the American Geophysical Union. - Geophysical Research Letters

A Guide to the Climate Debate

Hydrosphere	A hydrosphere is the total amount of water on a planet. The hydrosphere includes water that is on the surface of the planet, underground, and in the air. A planet's hydrosphere can be liquid, vapor, or ice. On Earth, liquid water exists on the surface in the form of oceans, lakes, and rivers.
IPCC	The Intergovernmental Panel on Climate Change was established in 1988 by the United Nations under the co-sponsorship of the World Meteorological Organisation (WMO) and the United Nations Environmental Programme (UNEP), with a charter that directs it to assess peer-reviewed research relevant to the understanding of the risk of human induced climate change.
Lithosphere	A lithosphere is the rigid, outermost rocky shell of a terrestrial planet or natural satellite. On Earth, it is composed of the crust and the lithospheric mantle, the topmost portion of the upper mantle that behaves elastically on time scales of up to thousands of years or more.
MCCIP	Marine Climate Change Impacts Partnership
Milankovitch cycles	Changes in the earth's elliptical annual orbit around the sun are known as the Milankovitch Cycles. These cycles affect the eccentricity, tilt and precession of the earth's orbit and hence distance from the sun and this affects the amount of energy hitting the earth.
NASA	The National Aeronautics and Space Administration is an independent agency of the U.S. federal government responsible for the civil space program, aeronautics research, and space research
Net Zero	Policy of the UK and other governments with the target of completely negating the amount of greenhouse gases produced by human activities and when balanced with implemented methods of absorbing carbon dioxide from the atmosphere, to achieve net zero emissions. The UK target is to achieve net zero by 2050.
Net Zero Watch	UK based Sceptic blog and website founded by Nigel Lawson, the former UK Chancellor. The site claims; *"Net Zero Watch highlights the serious implications of expensive and poorly considered climate change policies"*.

A Guide to the Climate Debate

NGO	A non-governmental organisation (NGO) is a group or organisation that functions independently of any government with the objective of improving social conditions. Environmental NGOs include Greenpeace, World Wildlife Fund, Friends of the Earth and Sierra Club
NOAA	National Oceanic and Atmospheric Administration, U.S. Department of Commerce
Paleoclimatology	alt: Palaeoclimatology – Is the scientific study of climates predating the invention of meteorological instruments when no direct measurement data were available.
Peer review	Peer review is an academic term for quality control. Each article published in a peer-reviewed journal was closely examined by a panel of reviewers who are experts on the article's topic (that is, the author's professional peers...hence the term peer review). The reviewers assess the author's proper use of research methods, the significance of the paper's contribution to the existing literature, and check on the authors' works on the topic in any discussions or mentions in citations. Papers published in these journals are expert-approved...and the most authoritative sources of information for college-level research papers.
Real Climate	A Warmist blog website; https://www.realclimate.org/
SAR	IPCC Second Assessment Report
Sceptic (climate sceptic)	Individual or organisation that challenges one or more aspect of anthropogenic climate change. Sceptics are often referred to as Deniers
Scientific Method	The scientific method is an empirical method for acquiring knowledge that has characterized the development of science since at least the 17th century. The scientific method involves careful observation coupled with rigorous scepticism, because cognitive assumptions can distort the interpretation of the observation.
TAR	IPCC Third Assessment Report
Temperature measurement	Temperature is a physical quantity that quantitatively expresses the attribute of hotness or coldness. The most common scales are Celsius scale with the unit symbol °C (formerly called centigrade), the Fahrenheit (°F), and the Kelvin scale (°K), the latter is used predominantly for scientific purposes. Absolute zero kelvin or -273.15°C is the lowest point with a 1°C change the same as a 1°K

A Guide to the Climate Debate

	change. The Celsius scale is used within this evidence review. https://en.wikipedia.org/wiki/Temperature
TED (talks)	Technology, Entertainment and Design
TSI	Total Solar Irradiance
UEA	University of East Anglia
UN	United Nations
UHI	Urban Heat Island effect. The warming of urban areas due to human activity compared to rural areas.
Warmist	Individual or organisation that supports anthropogenic climate change
Watts Up With That	Sceptic blog website edited by Anthony Watts; https://wattsupwiththat.com/

2 - References

The reference list includes links to web sites and scientific papers.

Ref	Title, Author, Description
1	Climate the movie, Documentary movie by Martin Durkin; https://www.climatethemovie.net/
2	An inconvenient truth, Documentary movie by Al Gore. Watch An Inconvenient Truth | Prime Video (amazon.co.uk)
3	United Nations Climate web page. https://www.un.org/en/climatechange
4	Intergovernmental Panel on Climate Change (IPCC), web site https://www.ipcc.ch/about/
5	UK Met Office – What is climate, https://www.metoffice.gov.uk/weather/climate/climate-explained/what-is-climate
6	Central England Temperature (CET) record, History of Information web site; https://www.historyofinformation.com/detail.php?id=3537
7	Met Office Central England Temperature (CET) record, https://www.metoffice.gov.uk/research/climate/maps-and-data/cet-series
8	Teaching a Model-based Climatology Using Energy Balance Simulation, David Unwin 1981 Teaching a model-based climatology using energy balance simulation: Journal of Geography in Higher Education: Vol 5 , No 2 - Get Access (tandfonline.com)
9	Climate change indicators, US EPA web page - Climate Change Indicators: U.S. and Global Temperature | US EPA
10	Global temperatures, NASA Earth Observatory web page - https://earthobservatory.nasa.gov/world-of-change/global-temperatures
11	EPA Climate change indicators, USA Environmental Protection Agency (EPA), climate change indicators - https://www.epa.gov/climate-indicators/weather-climate
12	EPA high & low temperatures, USA EPA - Climate change indicators, high & low temperatures - https://www.epa.gov/climate-indicators/climate-change-indicators-high-and-low-temperatures
13	EPA Heat Waves, USA EPA - Climate change indicators, heat Waves - https://www.epa.gov/climate-indicators/climate-change-indicators-heat-waves

A Guide to the Climate Debate

14	EPA drought, USA EPA - Climate change indicators, drought - https://www.epa.gov/climate-indicators/climate-change-indicators-drought
15	EPA heavy precipitation, USA EPA - Climate change indicators, heavy precipitation - https://www.epa.gov/climate-indicators/climate-change-indicators-heavy-precipitation
16	EPA Global Precipitation, USA EPA – Global precipitation - Climate Change Indicators: U.S. and Global Precipitation	US EPA
17	EPA river flooding, USA EPA - Climate change indicators, River flooding - https://www.epa.gov/climate-indicators/climate-change-indicators-river-flooding
18	TORRO, The Tornado and Storm Research Organisation https://www.torro.org.uk/extremes
19	EPA tropical cyclones, USA EPA - Climate change indicators, tropical cyclones - https://www.epa.gov/climate-indicators/climate-change-indicators-tropical-cyclone-activity
20	Wildfires, Our World in Data, wildfires - https://ourworldindata.org/wildfires
21	Carbon cycle, Met Office, carbon cycle - Carbon cycle - Met Office
22	Global carbon cycle, Science Direct, global carbon cycle - https://www.sciencedirect.com/topics/earth-and-planetary-sciences/global-carbon-cycle#chapters-articles
23	Modified atmosphere packaging, Air Products, modified atmosphere packaging - https://www.airproducts.co.uk/applications/modified-atmosphere-packaging
24	CO_2 best practice guide, AHBD, best practice guide - https://horticulture.ahdb.org.uk/knowledge-library/co2-best-practice-guide-background
25	The carbon cycle, NASA, the carbon cycle - https://earthobservatory.nasa.gov/features/CarbonCycle
26	Met Office blog 2 May 2024, Met Office, official blog 2 May 2024 - https://blog.metoffice.gov.uk/2024/05/02/the-atlantic-meridional-overturning-circulation-in-a-changing-climate/
27	Ocean circulation report, Marine Climate Change Impacts Partnership (MCCIP), ocean circulation report - https://www.mccip.org.uk/ocean-circulation
28	Briefing 7, the carbon cycle, Royal Society briefing 7, the carbon cycle - https://royalsociety.org/-/media/policy/projects/climate-change-science-solutions/climate-science-solutions-carbon-cycle.pdf
29	A graphical history of atmospheric CO_2 levels over time, Earth.org, - A Graphical History of Atmospheric CO2 Levels Over Time	Earth.Org

A Guide to the Climate Debate

30	Carbon Dioxide, NASA, carbon dioxide - Carbon Dioxide	Vital Signs – Climate Change: Vital Signs of the Planet (nasa.gov)
31	Atmospheric Carbon Dioxide, NOAA Climate Change, atmospheric carbon dioxide - Climate Change: Atmospheric Carbon Dioxide	NOAA Climate.gov
32	The greenhouse effect, British Geological Survey, the greenhouse effect - The greenhouse effect - British Geological Survey (bgs.ac.uk)
33	Earths global energy budget, Trenberth et al. 2009 Earth's Global Energy Budget in: Bulletin of the American Meteorological Society Volume 90 Issue 3 (2009) (ametsoc.org)
34	How well do we understand and evaluate climate change feedback processes? Bony et al. 2006 How Well Do We Understand and Evaluate Climate Change Feedback Processes? in: Journal of Climate Volume 19 Issue 15 (2006) (ametsoc.org)
35	Environmental effects of increased atmospheric carbon dioxide, Robinson, Robinson & Soon, 2007, Environmental effects of increased atmospheric carbon dioxide - https://www.oism.org/pproject/s33p36.htm
36	What role has the Sun played in climate change in recent decades? Royal Society, what role has the Sun played in climate change in recent decades? - 4. What role has the Sun played in climate change in recent decades?	Royal Society
37	The Central England Temperature Series, Met Office, The Central England Temperature Series - The Central England Temperature series - Met Office
38	Solar Influences on Climate, Imperial College London, Solar influences on climate, 2011 by Prof. Joanna Haig - Solar-Influences-on-Climate---Grantham-BP-5.pdf (imperial.ac.uk)
39	IPCC underestimates the sun's role in climate change, Bas van Geel & Peter A Ziegler, 2013, University of Amsterdam Climate800BC-VanGeel2013.pdf (ancientportsantiques.com)
40	Ice cores and climate change, British Antarctic Survey, ice cores and climate change – Ice cores and climate change - British Antarctic Survey (bas.ac.uk)
41	Ice Cores, Climate Data Information, Ice cores - Ice Core Data for Antarctic and Arctic (climatedata.info)
42	Climate myths: Ice cores show CO_2 increases lag behind temperature rises, disproving the link to global warming, New Scientist, Climate myths: May 2007 by Michael Le Page and Catherine Brahic https://www.newscientist.com/article/dn11659-climate-myths-ice-cores-show-co2-increases-lag-behind-temperature-rises-disproving-the-link-to-global-warming/

A Guide to the Climate Debate

43	Tracking sea level rise and fall, NOAA, Tracking sea level rise and fall - Tracking sea level rise ... and fall	National Oceanic and Atmospheric Administration (noaa.gov)
44	NOAA Global Sea Level, NOAA, Global sea level - Climate Change: Global Sea Level	NOAA Climate.gov
45	NASA Key Indicators, Global Mean Sea Level, NASA Sea Level Change, Understanding Sea Level, Key indicators, Global Mean Sea Level - Global Mean Sea Level	Key Indicators – NASA Sea Level Change Portal
46	IPCC Sea Level Rise, Sea Level Rise — IPCC
47	Sea Level Rise, IPCC paper titled; Sea Level Rise by R Warrick and J Oerlemans - ipcc_far_wg_I_chapter_09.pdf
48	AR6 Sea Level Projection Tool, NASA IPCC - IPCC AR6 Sea Level Projection Tool – NASA Sea Level Change Portal
49	A primer on pH, NOAA, - A primer on pH (noaa.gov)
50	Ocean Acidification, Yale, - Yale Experts Explain Ocean Acidification	Yale Sustainability
51	What is a GCM? IPCC, - What is a GCM? (ipcc-data.org)
52	On the recovery from the Little Ice Age dated July 2010 by Syun-Ichi Akasofu - On_the_recovery_from_the_Little_Ice_Age (1).pdf
53	Validity of Climate Change Forecasting for Public Policy Decision Making dated October 2009 by Kesten Green, J Scott Armstrong & Willie Soon - Naiveclimate (1).pdf
54	Global-scale temperature patterns and climate forcing over the past six centuries, Nature, by Michael E Mann, Raymond S Bradley & Malcolm K Hughes dated April 1998 mbh98.pdf (psu.edu)
55	Northern hemisphere temperature during the past millennium: inferences, uncertainties, and limitations, GRL, by Michael E Mann, Raymond S Bradley & Malcolm K Hughes dated March 1999 - Northern hemisphere temperatures during the past millennium: Inferences, uncertainties, and limitations (wiley.com)
56	Global surface temperatures over the past two millennia, GRL, by Michael E Mann & Philip D Jones, dated 2003 – https://agupubs.onlinelibrary.wiley.com/doi/10.1029/2003GL017814
57	TAR Summary for Policymakers, IPCC Third Assessment report - spm.pdf (ipcc.ch)
58	Proxy climatic and environmental changes of the past 1000 years, Climate Research, by Willie Soon and Sallie Baliunas, dated January 2003 - https://www.int-res.com/articles/cr2003/23/c023p089.pdf

59	Corrections to the Mann et al. (1998) proxy data base and northern hemisphere temperature series, Energy & Environment, by Stephen McIntyre and Ross McKitrick, dated 2003 - https://climateaudit.org/wp-content/uploads/2005/09/mcintyre.mckitrick.2003.pdf
60	Hockey sticks, principle components, and spurious significance, Hockey sticks, principal components, and spurious significance - McIntyre - 2005 - Geophysical Research Letters - Wiley Online Library
61	AR4 Summary for Policymakers, IPCC Fourth Assessment report - https://www.ipcc.ch/site/assets/uploads/2018/02/ar4_syr_spm.pdf
62	Climate Research Unit, CRU, University of East Anglia - https://www.uea.ac.uk/groups-and-centres/climatic-research-unit
63	How did the obsession with decarbonization arise? Net Zero Watch article by Richard Lindzen, - How did the obsession with decarbonization arise? (netzerowatch.com)
64	IPCC History, - History — IPCC
65	United Nations Framework Convention on Climate Change, United Nations (UNFCCC) - UNFCCC
66	Sixth Assessment Report, IPCC, AR6 - AR6 Synthesis Report: Climate Change 2023 — IPCC
67	How do scientists know, Royal Society, how do scientists know - 2. How do scientists know that recent climate change is largely caused by human activities?	Royal Society
68	Royal Society news 12 March 2018, Royal Society - The Royal Society joins Commonwealth science academies to call for action on climate change	Royal Society
69	DeSmog,blog site - News - DeSmog
70	Skeptical Science, blog site - About Skeptical Science
71	Real Climate, The leading Warmist scientific blog site – RealClimate: Frontpage
72	Climate Audit, is a Sceptic blog site - Climate Audit
73	Watts Up With That, is a climate Sceptic blog site - Watts Up With That? – The world's most viewed site on global warming and climate change
74	Net Zero Watch, is a UK Sceptic website - Net Zero Watch
75	Review of the movie 'Climate the Movie' by Net Zero Watch - Climate, the Movie: A Review (netzerowatch.com)
76	Why renewables can't save the planet, TED Talk by Michael Shellenberger - https://www.youtube.com/watch?v=N-yALPEpV4w
77	Explore Long-Term Strategies (LTS), Climate Watch website page - https://www.climatewatchdata.org/lts-explore?indicator=lts_submission

A Guide to the Climate Debate

78	Halfway between Kyoto and 2050, Fraser Institute paper by Vaclav Smil 2024 - https://www.fraserinstitute.org/sites/default/files/halfway-between-kyoto-and-2050.pdf
79	Understanding carbon capture and storage, British Geological Survey web page: - https://www.bgs.ac.uk/discovering-geology/climate-change/carbon-capture-and-storage/
80	CO_2 emissions, Our World in data, CO_2 emission web page - CO_2 emissions - Our World in Data
81	An energy sector roadmap to carbon neutrality in China, International Energy Agency report 2021 An energy sector roadmap to carbon neutrality in China – Analysis - IEA
82	Sustainable Development Goals, United Nations Sustainable Development Goals - https://sdgs.un.org/goals

3 - Reading List

Ref	Book Title & Description

A   **Climate: The Counter Consensus**
Professor Robert M. Carter; Published 2010
The counter-consensus to quasi-scientific hype and induced panic on climate change is at last assembling. The argument is not in the first place as to whether or not climate change has been taking place, but whether any recent warming of the planet is appreciably due to human activity and how harmful it will prove. Tom Stacey, in his eloquent and provocative introduction, investigates our tendency to ascribe this and other perceived planetary crises to some inherent fault in ourselves, be it original sin or a basic moral failing.

Climate Change goes on to examine, with thoroughness and impartial expertise, the so-called facts of global warming that are churned out and unquestioningly accepted, while the scientific and media establishments stifle or deride any legitimate expression of an opposing view. In doing so, the book typifies the mission of Independent Minds to replace political correctness and received wisdom with common sense and rational analysis.

Book description from Amazon; https://www.amazon.co.uk/Climate-Counter-consensus-Professor-Robert-Carter/dp/1906768293

B   **Climate: The Great Delusion**
Christian Gerondeau; Published 2010
The author argues that the complex and formidably expensive measures undertaken by the international community to combat climate change are unworkable and unnecessary. Specifically, these measures: will have no effect on climate change, if indeed it is actually occurring; will be nullified by the actions of China and India; make no sense in the light of our hydro-carbon sources being exhausted within the coming century and will come to be viewed as an incredible response to global delusion and harmless climate shifts. Under its French title of 'CO$_2$, Un Mythe Planetaire', this book - thoroughly updated for its English edition- has sold over 5,000 copies.

A Guide to the Climate Debate

Book description from Amazon; https://www.amazon.co.uk/Climate-Great-Delusion-Independent-Minds/dp/1906768412

C **Climate Change Isn't Everything**
Mike Hulme; Published 2023
The changing climate poses serious dangers to human and non-human life alike, though perhaps the most urgent danger is one we hear very little about: the rise of climatism. Too many social, political and ecological problems facing the world today – from the Russian invasion of Ukraine to the management of wildfires – quickly become climatized, explained with reference to 'a change in the climate'. When complex political and ethical challenges are so narrowly framed, arresting climate change is sold as the supreme political challenge of our time and everything else becomes subservient to this one goal.

In this far-sighted analysis, Mike Hulme reveals how climatism has taken hold in recent years, becoming so pervasive and embedded in public life that it is increasingly hard to resist it without being written off as a climate denier. He confronts this dangerously myopic view that reduces the condition of the world to the fate of global temperature or the atmospheric concentration of carbon dioxide to the detriment of tackling serious issues as varied as poverty, liberty, biodiversity loss, inequality and international diplomacy. We must not live as though climate alone determines our present and our future.

Book description from Amazon; https://www.amazon.co.uk/Climate-Change-isnt-Everything-Liberating/dp/1509556168

D **Hiding the Decline**
Andrew W Montford; Published 2012
In November 2009, hundreds of emails from the Climatic Research Unit at the University of East Anglia were released onto the internet. The messages, sent between some of the world's most prominent climatologists revealed an extraordinary array of malpractice, with scientists manipulating data, breaching freedom of information laws and trying to crush dissenting views.

Hiding the Decline is the definitive history of the Climategate affair, tracing the story back to its roots in the struggle over the notorious Hockey Stick graph, reviewing the explosive revelation of the emails themselves and then examining in forensic detail the cover-ups that followed.

Including important new information about how the malfeasance of those involved in Climategate was whitewashed by the scientific establishment, Hiding the Decline is a both an important historical record and an entertaining story in its own right.

Book description from Amazon; https://www.amazon.co.uk/Hiding-Decline-W-Montford/dp/1475293364

E   **The Hockey Stick Illusion**
Andrew W Montford; Published 2010
From Steve McIntyre's earliest attempts to reproduce Michael Mann's Hockey Stick graph, to the explosive publication of his work and the launch of a congressional inquiry, The Hockey Stick Illusion is a remarkable tale of scientific misconduct and amateur sleuthing. It explains the complex science of this most controversial of temperature reconstructions in layperson's language and lays bare the remarkable extent to which climatologists have been willing to break their own rules in order to defend climate science's most famous finding.

Book description from Amazon; https://www.amazon.co.uk/Illusion-Climategate-Corruption-Science-Independent/dp/1906768358

F   **The Little Ice Age – how climate made history 1300 - 1850**
Brian Fagan; Published 2000
The Little Ice Age tells the story of the turbulent, unpredictable, and often very cold years of modern European history, how this altered climate affected historical events, and what it means for today's global warming. Building on research that has only recently confirmed that the world endured a 500 year cold snap, renowned archaeologist Brian Fagan shows how the increasing cold influenced familiar events from Norse exploration to the settlement of North America to the Industrial Revolution. This is a fascinating book for anyone interested in history, climate, and how they interact.

Book description from Amazon; https://www.amazon.co.uk/Little-Ice-Age-Climate-1300-1850/dp/0465022723

Printed in Great Britain
by Amazon